Hillsboro Public Library
Hillsboro, OR
A member of Washington County
COOPERATIVE LIBRARY SERVICES

Black Women Will Save the World

Black Women Will Save the World

AN ANTHEM

April Ryan

AMISTAD
An Imprint of HarperCollinsPublishers

BLACK WOMEN WILL SAVE THE WORLD. Copyright © 2022 by April Ryan. All rights reserved. Printed in the United States of America. No part of this book may be used or reproduced in any manner whatsoever without written permission except in the case of brief quotations embodied in critical articles and reviews. For information, address HarperCollins Publishers, 195 Broadway, New York, NY 10007.

HarperCollins books may be purchased for educational, business, or sales promotional use. For information, please email the Special Markets Department at SPsales@harpercollins.com.

FIRST EDITION

Designed by Elina Cohen
Illustrations on pp. iii, 21, 81, 127 © (Plawarn)/Shutterstock
Illustrations on pp. 1, 23, 48, 64, 83, 99, 116, 129, 151 © (oksanka007) /Shutterstock

Library of Congress Cataloging-in-Publication Data has been applied for.

ISBN 978-0-06-321019-6

22 23 24 25 26 LSC 10 9 8 7 6 5 4 3 2 1

To my late mother, Vivian; her mother, Etta; her mother, Ida; and her mother, Laura; and to the unnamed others who came before them and endured the atrocities of slavery to safely deliver the next generation.

As Maya Angelou penned in her poem "Still I Rise," "I am the dream and the hope of the slave." I descend from Nigerian, Ghanaian, Congolese, and Senegalese foremothers whose strength preserved the dream of a brighter future. This book is dedicated to each and every one of them. Their strength, and blood, flows through me and my daughters.

Contents

	Foreword by Cory Booker	*ix*
Preface	**Extraordinary Ordinary**	*xiii*
Introduction	**What's at Stake**	*1*

Section I

The Power
HOW WE LEAD AND WHY

Chapter 1	**The Superpower of Sisterhood**	23
Chapter 2	**How Not to Be Erased**	48
Chapter 3	**Walking the Tightrope**	64

Section II

The Price

WHAT WE ENDURE, HOW WE OVERCOME

Chapter 4	**Our Fight**	*83*
Chapter 5	**Our Sacrifice**	*99*
Chapter 6	**Our Voice**	*116*

Section III

The Promise

WHAT'S NEXT FOR US AND AMERICA

Chapter 7	**"A Little Child Shall Lead Them"**	*129*
Chapter 8	**Healing**	*151*
Chapter 9	**What's Next**	*163*
	Acknowledgments	*175*
	Notes	*177*

Foreword

BY SENATOR CORY BOOKER

I am who I am today in large part because of incredible Black women.

When I look back on my life, I don't need to look very far to see the many Black women whose extraordinary investment in me shaped me into the man I am today. When I reflect on US history, Black women have been some of the most extraordinary benders of the long arc of our moral history toward justice. Through their grit, guts, light, and love, they have strengthened this country even though this country didn't always love them back.

My mother, Carolyn, was my first hero and the ultimate role model across all of my life. She was active in the civil rights movement and, in 1963, helped to organize the March on Washington. She raised my brother and I to understand that the privileges we enjoyed as Americans were paid for by the sacrifice and struggle of others. She imparted to me the facts: you can't pay back the blessings you receive, but you must through struggle pay them forward.

FOREWORD

After I finished law school, I decided to commit my life to the fight to make the American dream real for everyone. That commitment led me to Newark, to the neighborhood I call home to this day, where I joined with others to fight for people whose worth and dignity go too often unseen and unappreciated. There I met extraordinary Black women who so often led communities, held them together, and empowered them through the force of their love. One of those great women was Ms. Virginia Jones, the tenant president of the housing project where I lived. She gave me a lesson that I will never forget.

One day, not long after I first met her, we stood on Dr. Martin Luther King Jr. Boulevard, where she asked me to "take a look around this neighborhood and tell me what you see."

My response was simple. "I see projects. I see abandoned buildings. There is a bodega down the block…"

She was obviously angry and disappointed. "You can't help me. You can't help me."

"What do you mean?" I replied, confused.

"The world you see outside of you is a reflection of what you have inside of you," she started. "If you only see problems and darkness and despair, then that's all there's ever gonna be. But if you're stubborn, and every time you open your eyes, you see love, you see the face of God, then you can help me."

Ms. Jones transformed my perspective on the world. She knew the worth of her neighborhood, the potential of its people. And she never gave up. She had a defiant love for her people and community, despite profound personal pain and tragedy.

My personal experience is by no means unique. Our nation's past is filled with incredible, capable Black women who have left a lasting mark on our communities and our country, some whose names we know and many more who are forgotten by history.

Despite persistent efforts to hold them back, despite carrying scars from being qualified yet still so often denied, Black women continue to defiantly love our country. We are all better off because of it.

Today, Black women are increasingly breaking through glass ceilings and ascending to many of the positions of political power, prestige, and prominence that they have more than earned, from Kamala Harris to Ketanji Brown Jackson. Black women collectively are also showing their might at the voting booth, ranking among the most active (and important) voting blocs in the United States.

April Ryan has firsthand experience with both witnessing and playing a huge role in Black women's long overdue ascendance, chronicling the halls of power in Washington for over twenty-five years. She has navigated many of the obstacles that have stood in the way of so many Black women and, in the process, broken so many barriers herself.

April is filled with extraordinary toughness and determination. She speaks to our common values, the enduring and urgent need for truth tellers, and how often the most difficult challenges bring out the best in who we are. She is a living example of how the fire of adversity reveals and forges greatness. And she loves this country.

That is why she's such a fitting person to share this story—the story of Black women's rise.

My hope is that these pages will inspire all people to carry forward the work that Black women in America have been carrying for years. The story of Black women's ascendance is a great American story and should help renew, reaffirm, and reignite our commitment to the unfinished business of this country.

PREFACE

Extraordinary Ordinary

February 25, 2022, 2 p.m., in the Cross Hall of the White House. The world watched to mark *herstory*: the very first nomination of a Black woman to the United States Supreme Court.

I was at home in icy Baltimore, working on various stories to commemorate the historic nomination. I was also trying to wrap my mind around the fact that a colleague of mine—a white man—was given the story to break.

As I gazed upon the televised images of Judge Ketanji Brown Jackson and Vice President Kamala Harris—two powerful, superbad Black women standing behind the president of the United States—I thought about the scores of Black women who have fought, are fighting, and will fight to break into the ultimate power structure, a feat that had eluded us for so long. I fought back tears. And anger. Tears for the triumph. Anger for its price.

President Joseph Biden said of Judge Jackson that she was chosen

PREFACE

for her "legacy of excellence and decency." He called out her "strong moral compass" and said he believed that her ability "to stand up for what's right" was what the court needed today.

The optics—Judge Jackson standing next to Vice President Harris just a year after Harris's historic ascension to the second-highest office of the land—showed the entire world that Black women were no longer just heard but *seen*.

As she savored the moment, Judge Jackson remarked, "I thank God for delivering me to this point. One can only come this far by faith."

Indeed.

Today's generations of Black women have arrived here if only by faith—a faith that we keep and a faith kept by our ancestors in spite of the unfathomable pain, torment, and exclusion in centuries past.

All my life, I have known and embraced—even if, at times, unconsciously—the undeniable fact *that Black women will save the world*.

How could I not believe that essential truth? I did not have to go far.

As a child, I watched my amazing mother move mountains with grace and conviction. She raised me from the cradle to be strong and wise. She showed me how. My mother always held her ground: She created space for me to grow up safe and secure in who I was, while lifting up the young people in our community—a community of all ages, races, and genders—by helping them excel and achieve their best life. She was, in every sense of the word, *magic*.

For Black women reading this, that's a familiar sentiment. *Magic*. Well, we are magic. Juggling it all and helping everyone, including and especially the times when no one notices. Black women: we make the extraordinary *ordinary*.

This has been unfailingly true throughout our history. As I grew

PREFACE

older and learned about our past, the remarkable leadership qualities of Black women circled back again, and again, and again. Black women have played an extraordinary and largely unacknowledged role in the arc of our country.

In the Revolutionary War, Black women were wildly effective spies, often working from behind enemy lines, unseen and unappreciated as house and field workers, and often enslaved, fighting for that nascent idea of America while keeping the faith that one day we might be free too—that one day we might arrive.

Throughout the nineteenth century, as we continued to toil and drive forward the nation's economy in the North and the South, we organized, invisible to most, a movement for the abolition of slavery, meeting in churches and living rooms and community gatherings. We served as conductors on the Underground Railroad, delivering one soul after another to freedom.

Our efforts roused the nation's conscience and when America plunged into Civil War, we were among the first to volunteer as spies, soldiers, and nurses—even though our freedom, fully and truly, wasn't on the docket.

We have always been the first to answer the call. And through it all, we have kept the faith.

In the long decades that followed the Civil War's end, a time marked by political, social, and economic backlash characterized by the laws of Jim Crow, segregation, economic subjugation, and violent lynchings, we never stopped pushing and fighting for what is right. The women's right to vote. The right to sit at the front of the bus. The right to a quality education. The right to our bodies. The right to fully participate in this American Experiment.

And through it all, we have kept the faith.

PREFACE

When Judge Jackson said, "I thank God for delivering me to this point. One can only come this far by faith," heavy tears fell down my cheeks.

I knew what she meant. *We knew*.

After all, Black women make the extraordinary ordinary. We drive change in a society that wasn't built for us *in order to make it better for all of us*.

We do this, instinctually, because it is always what needs to be done.

Few of us will reach the acclaim and historic heights of Judge Jackson or Vice President Harris, but the contributions of all Black women matter. We deserve acknowledgment because invariably it is our efforts—our faith—that propel the American project forward. For me, this book is a love letter to Black women everywhere. *I see you*. And I thank you.

Look around your community as we fight through this horrific pandemic. Contemplate the caretakers, the frontline workers, and, perhaps, even your local elected officials—the Black women of our society stepping up to serve, to help, to lend a hand when it's needed most. *Again*.

All of us have been affected by the extraordinary efforts of Black women during this historic time of trial. And we all have stories and observations about why Black women matter.

I think about my cousin Melba, a restaurateur in Harlem, and the ways she fed thousands of hungry people and hundreds of first responders, day after day, week after week, "pay only what you can," when the COVID-19 vaccine was all but a pipe dream, just a protein on a screen.

PREFACE

I reflect on the creative vision of our role models like Beyoncé, who belts out, "Who run the world? Girls!"—a national anthem for Black women that reminds us that "my persuasion can build a nation." An anthem repeated over and over as a new generation of Black girls internalize their greatness and see it reinforced in leaders like Karine Jean-Pierre, President Biden's press secretary as of spring 2022. Jean-Pierre speaks from the presidential lectern for all the world to see, wisely guiding public opinion as the nation navigates war overseas and a deadly virus here at home.

Watching Jean-Pierre from the lectern, I am elated: there isn't a day that goes by when I do not drink it all in. *One can only come this far by faith.*

I imagine, with awe, the African women in London who are raising money for the Black migrants fleeing Ukraine. They remind me that we are a global movement—Black women—fighting for what's right wherever we find ourselves.

In the quiet moments, I remember the heartache of 2016: the painful elevation of bigotry, the threat to democracy. I can still feel in my gut that heavy sense of loss that consumed millions of Americans—and our hopelessness. Our despair. And then I remember that we, Black women, fought back to pull the country back from the brink. We won the special election in Alabama. And then the United States Senate races in Georgia. And then the presidency.

Which brings us to this moment. *Right now.* Unlike in other chapters of American history, our presence is not just felt, our voices are not just heard, but *we are seen. We are powerful. We have arrived.*

The extraordinary made ordinary. That's what we do. All in a day's work.

PREFACE

The only thing missing now is a *conversation* about who we are, how we lead, the price we pay, and what's next for us and a country forever changed by our ascent. I have had the privilege to talk to Black women and study the enormous contributions of Black women today and stretching back through time. My fervent hope is you will find these women as inspiring as I have. Most important, I hope their example pushes you, indeed all of us, to elevate and exalt Black women as the sheroes and world shapers they are and always have been.

We are undeniable now. It's time.

April Ryan, 2022

Black Women Will Save the World

INTRODUCTION

What's at Stake

> *Black women prove to be the backbone of American democracy.*
>
> Headline for an article written by Shelby Stewart, in the *Houston Chronicle*, January 6, 2021

OAK BLUFFS, MARTHA'S VINEYARD

I unapologetically believe we need to take time for ourselves. Since 1999, I have "decamped" to Oak Bluffs on Martha's Vineyard for vacation.

Oak Bluffs represents more than respite. The bucolic, warm, lush Massachusetts island community offers a chance to reconnect with traditions that have held Black women together dating back to the trials and tribulations of Jim Crow. This historic African American summer resort community, founded in the nineteenth century, was our getaway during the fight for civil rights in the 1950s and 1960s. Decades later, it remains a powerful place of rest. We commune spiritually at the Union Chapel, built in the Victorian era as a nonsectarian place of worship. On Sundays, the chapel's beautiful architecture reverberates with moving sermons and uplifting song.

During the summer, Oak Bluffs swells with joyous college and university reunions as Black sorority sisters, like my beloved Delta Sigma Theta Sorority Incorporated sisters, gather to celebrate and reflect. Laughter and trash talk can be heard in between bites of lobster rolls, handmade fudge, and soft-serve ice cream. An ocean breeze that cools and soothes swirls across the island. Rows of gingerbread houses transport you, instantly, to another time.

Whenever I am on the island, I always delight in the company of other Black women. Oftentimes, that company is kept in silence. On one recent trip, I sat next to a woman who was, like me, lost in the sounds of the beach, watching the "polar bears"—those brave souls who sojourn to the Inkwell Beach to work out in the not-so-cold water. We were both taking it all in: our people, the water, the waves, and the sand between our toes as we sat on a park bench, just watching. The woman on the bench did not look familiar to me. But we were both so absorbed in the oceanic rituals unfolding that it didn't matter. We were simply present—two women basking in the moment.

The next day I remembered that Nikole Hannah-Jones, the Pulitzer Prize–winning journalist who created the landmark *1619 Project* for the *New York Times,* was on the island. I messaged her on Twitter asking if we could meet up. She messaged right back.

"Oh my," she started, "I think I might have been sitting right next to you at the Inkwell yesterday. Is your hair in cornrows?"

I *was* in cornrows. Her signature, bright, beautiful red hair had not been showing. I smiled. We had spent the morning together, in tranquility, and didn't even know it. All that mattered was that we had the chance to be in the moment—a chance to breathe, an opportunity to appreciate our people, our joy, and our strength.

My mission on this vacation, as it is with every trip to Oak Bluffs, was introspection. I spent August of 2021 basking in my Black womanhood, reflecting on our wins and losses from the last year: President Biden's victory and the election of our first Black and female vice president; the insanity of January 6, when the forces of hate and division attempted to turn back the clock, once again, on American progress; the turbulence that follows the ascension of a new president—from troop withdrawal from Afghanistan to the twin epidemics of COVID-19 and the disinformation that keeps us from defeating it; and the "long, twilight struggle" for voting rights and police reform.[1] And on, and on, and on. However, through it all, Black women remain vigilant and essential—fighting the fight and, in many cases, leading the charge.

In this moment, the world is *full*—full of fight and struggle, full of hope and aspiration, full of anxiety and uncertainty. In this moment, Americans of all stripes and persuasions are sense making, trying to figure out how we escape our present, dismal circumstances to achieve something brighter, more hopeful, and fundamentally fair.

In that sense making, we are all looking for heroes. For some Americans, their hero is ensconced in a palace of his own making in Florida (and, for my part, let's hope he stays there). Many others have focused their attention on the Black women who are guiding America through one of her darkest, most difficult chapters. *Black women will save the world*. To my core, I believe in that statement. Black women have saved the world before and are doing so now.

Today Black women walk the halls of Congress in historic numbers. We are leading on the front lines of activism. In the White House press corps and now in the White House itself, we offer America her moral compass as she decides between two visions for the future: one that is broad, inclusive, and fair, and another that is undemocratic, dangerous, and hateful. Silently, and sometimes loudly, in the spotlight but more often out of it altogether, Black women are the ones doing everything we can—I am paraphrasing the late, great Black writer Langston Hughes—to: "Let America be America again. Let it be the dream it used to be. . . . O, let my land be a land where Liberty is crowned with no false patriotic wreath, but opportunity is real, and life is free, equality is in the air we breathe."[2]

For Black women, our moment is here. And in the history of the United States, the reach, influence, and power of Black women are unprecedented. The inequity that follows four hundred years of sexism and racism hasn't been erased, but Black women, as a category, are triumphing. And we are ascending at the precise moment when America needs our leadership. *Black Women Will Save the World* chronicles, in real time, the heroic efforts of Black women, across industries and professions, to move the needle on questions of justice and fairness. This book is also a candid portrait of leadership—the disappointments, the frustrations, and, indeed, the costs of leading America during one of her darkest hours.

Black women are triumphing, but we also continue to pay a profound price for our personhood *and* our leadership. This book is an unflinching look at what it's like to be in the room with some of the most formidable Black women in leadership in all the world as they navigate a world that wasn't built for them but needs them more than ever.

As COVID-19 swept through America, claiming the lives of a million Americans and creating new and unique economic and political threats, it was Vice President Harris, a Black woman, who was driving America's response—to quote President Biden—to "build back better."

The vice president is now juggling the president's most sensitive portfolios, including a global migration crisis and the imperative to advance racial justice across the whole of government in the wake of the deaths of George Floyd, Breonna Taylor, and Ahmaud Arbery.

It was California Representative Maxine Waters, a Black woman, who sounded the alarm on January 6, and it was ten Black women in the United States Congress who went to jail in protest against the corruption of our elections and destruction of our voting rights. It was Stacey Abrams, a Black woman in Georgia, who delivered a decisive victory to presidential candidate Biden and the Democrats in decisive, can't-lose elections.

It was Dr. Kizzmekia Corbett, a Black woman, who led the development of the Moderna COVID-19 vaccine. Following the murder of George Floyd, as the country fell headlong into a culture war that bordered on civil war, it was activist Tamika Mallory, a Black woman, who set the movement for #BreonnaTaylor in motion, giving Taylor's story a powerful political life. Mallory compelled us to #SayHerName.

It's a tall order, saving America, and yet here we are.

When my feet touch the coarse surface of the Inkwell, the shards of seashells, which are sharp and painful for tender feet, remind me of the irony of our position: For centuries, Black women have been relegated to the most beautiful *and the most painful* beach on the island. And the generations who navigated its prickly shoreline have advanced a struggle for fairness despite the fact that the country doesn't always love us back. When Black women fight, we fight for everyone. We fight for our families, our communities, and, yes, our country.

Our struggle has never been more important to the future of the country.

Hate is on the rise. The virus rages on. Our Democracy is imperiled.

And the Threat is barely contained. And yet, Black women are ascendant. We are shattering glass ceilings, wielding power and influence at levels never before seen in this country. The culture is starting to recognize and acknowledge our power. "Black women saved the Democrats. Don't make us do it again," declared the *Washington Post* in the run-up to the 2020 general election. "Last night in Georgia, Black women saved democracy," exclaimed the Brookings Institution after President Donald Trump's defeat in November 2020. "Black women prove to be the backbone of American democracy," the *Houston Chronicle* underscored after the January 6 insurrection.

It is gratifying to see Black women's stock rise—finally—as evidenced by our increasingly positive portrayal in the media, by the public, and in the culture. And yet, for every exclamation that "Black girls are magic," we must also grapple with the hard realities: Black women still absorb a disproportionate share of this country's bias, discrimination, and material hardship. Black women earn an average of $5,500 less per year, and experience higher unemployment and higher poverty rates than what is the average for women. We are more likely to be the head of a household than our white female counterparts, effectively supporting more dependents with fewer resources. We live in neighborhoods that are more racially segregated and have lower property values than those of our white counterparts. The racist practice of "redlining," or residential segregation, has isolated many of us from opportunities in the job market and constrained our access to quality health care, capital, and education. We live in poorer neighborhoods and communities; residential segregation damns us to shorter, less healthy lives, as has been exhaustively demonstrated by scholars and academics.[3]

In formal leadership positions, we are grossly underestimated. Although we've made progress among the ranks of political leadership, there have only been three Black women to serve as CEO in a Fortune 500 company. In our other roles at work and in professional spaces, our ideas are taken less seriously or ignored altogether. Our contributions are more likely to be taken for granted or co-opted completely as colleagues take credit for our work while benefiting, often without attribution or affirmation, from our lived experience. We experience micro- and macro-aggressions inside and outside the home. We are harassed far more frequently, and yet, we are the "forgotten" survivors of sexual assault.[4]

In other words, we have a *long* way to go. *Black Women Will Save the World* will also give language to the everyday struggles of Black women and the ongoing bias and discrimination that underscore that struggle. This is important. As Black women increase in our power and profile, we must never forget that the basic facts of our lives are often worse than any other demographic group in America. This must change. And with your help, it will. Black women are the Hope, and our work drives progress—but we deserve allyship and solidarity too. We are worthy.

None of that is to say that Black women are not used to the underdog fight. We remain perennially marginalized and discounted, if not counted out altogether. And while many would consider that a weakness, Black women have transformed the experience of marginalization into a superpower. When confronted with injustice, we fight back and we fight back for everyone. For fairness. For justice. We are America's unheralded Sheroes. We wear the invisible cape and fly through the air to scan the horizon for injustice. We always find a way to save the day.

What's missing is *your* voice. My ultimate goal for this book is to inspire our country to celebrate and elevate Black women—our history, our heroism, our hurt, and our Hope. Much like what the Brothers Grimm did with their folktales long ago, it's the time to bring Black women into American folklore so that our example can inspire the next generation and shape the culture. Through a series of profiles, editorials, and my personal reflections as a journalist on the front lines of political reporting, this book tells the story of all the ways that Black women are saving America during one of her darkest chapters. Hate is on the rise in America, *but so is Hope.* Together, we can tell and share a new story of what's possible across the land we love.

OUR GENERATION'S NEXT FIGHT: HATE AND EXTREMISM

In 2017, we watched in horror as Charlottesville, a small, quaint Virginia college town, descended into chaos. In response to the city's decision to remove a statue of Confederate General Robert E. Lee, the Proud Boys, Oath Keepers, and other far right groups had traveled there from all over the country for the Unite the Right rally. Violence erupted when people protesting against the rally and the rally participants clashed.

On the second day of the rally, James Alex Fields Jr., a 20-year-old, self-proclaimed neo-Nazi in from Ohio, slammed his car into anti-rally protesters. Tragically, Heather Heyer, a young and courageous white woman who was one of the peaceful marchers, later died from her injuries. But what did we hear from the president then? "There are fine people on both sides," President Donald Trump proclaimed at a press conference a few days after the rally. Words fail me.

The FBI would eventually declare the vehicle attack an act of domestic terrorism. Since then, the FBI and the country's national security apparatus have declared white supremacy to be the country's most dangerous domestic terrorist threat. Reports of racist threats have skyrocketed to levels unseen since Reconstruction, a period that immediately followed the end of the Civil War and the abolition of slavery in 1865 and ended in 1877, when federal troops withdrew from the South.

Charlottesville's peaceful, beautiful Main Street, which I had visited just months before, forever changed after the terrorist attack. Long a *noun* that connoted bucolic beauty and serenity, Charlottesville now suggests a nasty *verb*: the action of hate.

Four years later, in 2021, I reconnected with my friend, a jewelry shop owner on Charlottesville's Main Street, to recall the events of August 2017. She vividly remembered the spectacle of the protests, her disgust at the racial animus that visited Charlottesville, *the fear* that permeated that hot August day, and the violence that ultimately claimed one woman's life, left dozens more injured, and changed thousands forever. "All of my life, I have been a Jewish woman in Charlottesville, and I have never feared for my life until that day," she shared with me, reflecting on the aftermath of the attack.

This friend is one of millions of Jewish people grappling with an unprecedented rise in anti-Semitism across the country, catalyzed by President Trump's divisive politics. In 2020, the Anti-Defamation League (the premier organization that stands against anti-Semitism) recorded the highest levels of anti-Semitic activities since the organization started tracking incidents forty years ago. She discussed how the protesters courageously protected small businesses just like hers from further destruction as the Unite the Right white supremacists lashed out at people and property alike to wreak havoc and destruction.

"I don't care if they're called antifa; they looked like young and committed college students to me. The police just stood by and watched it all unfold. It was unbelievable."

My friend recalled how the white supremacists, decked out in camouflage and carrying automatic weapons, prepared for their rally in a nearby city parking garage where, once again, police looked the other way. "Honestly, it was a lot like January 6 when police let them [the white supremacists] overtake the Capitol. For me, the two events are almost identical. I can't unsee that."

Brave patriots, many of them young people, prevented white supremacists from destroying Charlottesville, literally and figuratively, that fateful day. We should never forget the young white woman, Heather Heyer, who gave her life to stand up against hate in the name of equality and dignity for all. Allies have a long and important history in the struggle for human rights in the United States. Heather Heyer deserves our tribute and our remembrance.

Despite the gravity of the events in Charlottesville, many politicians at all levels of government have downplayed the deadly acts of hate that marred Charlottesville and stained the character of the country. As a response to the events, some lawmakers are even pushing anti-protest laws in order to give immunity to someone who strikes protesters with their vehicle. Much like the events of January 6, the political right has rationalized the bad behavior of a growing extremist element in American society. With each breach, that element grows stronger as the core of our country buckles under the pressures of bigotry and disinformation.

We live in dangerous times. American Democracy is under assault. Her chief defender, the Black woman, too often fights alone.

And *this*, my dear reader, is the point of *this* book.

In the twenty-first century, Black women have played a vital role in defending America. From politics and the media and journalism to the worlds of community and community activism, Black women have served as one of the final bulwarks against an epidemic of hate that threatens to overtake the country and change her fundamental character. We do so in an environment no less hostile than what our ancestors confronted. Black women today are as discounted and underestimated as we were when Shirley Chisholm announced her historic race to be the first Black women elected to Congress in 1968 and then for president in 1972, or when abolitionist and women's rights activist Sojourner Truth visited President Abraham Lincoln in the White House in 1864.

Nevertheless, the "double whammy" Chisholm pointed out of being Black and female in America has not deterred Black women from protecting the country from the most dangerous escalation of hate since the Civil War. Starting with 9/11, continuing through the time of President Barack Obama, and culminating in the presidency of Trump, America has experienced a steady increase in *hate*. That trend responds to America's growing multiculturalism and its place in an increasingly globalized world. The perceived ascendance of people of color, women, and gender minorities has ignited a backlash, largely from the "forgotten" white male whom President Trump fought so hard to elevate.

The backlash takes many shapes.

In some quarters, the backlash manifests itself as a conspiracy. For instance, for the duration of his candidacy and presidency, President

Obama endured an inordinate degree of racist hate. The White House press corps, of which I am a proud member, was complicit. We devoted an extreme amount of attention to President Obama's birth certificate, achieving two things: delegitimizing our first Black president and enfranchising the political hucksterism of Donald Trump (the loudest and most aggressive progenitor of the lie that Obama was not born in the United States).

The backlash is channeled through incessant propaganda and the larger War on Truth. Part of my role as a responsible and ethical journalist is to present the facts to the American people. Whether it is reporting on the COVID-19 vaccine, the assault on voting rights, or the integrity of our elections, part of my job is to help set the record straight through the standard question-and-answer reporter dance and to defend Truth from her many and motivated enemies.

Most dangerously, the backlash is violent, even insurrectionist. The events of Charlottesville and later of January 6, 2021, saw America grappling with the scourge of white supremacy at the highest levels of government. Despite the magnitude of these transgressions, many Americans are fighting to "forget" the trauma of those events by rewriting history in real time. Much like the Confederate monuments that came to punctuate the landscape of the Deep South after the Civil War and during the various movements for civil rights, agents of hate will stop at nothing to normalize their despicable behavior.

We Black women––politicians, journalists, activists, mothers, daughters, and sisters––will not let them. There is too much on the line.

While the quest for justice is never-ending, some chapters in the American story pose greater risks to freedom than others. America successfully launched a revolution, endured a civil war, and won two world wars. The nation has marched for civil rights and has stared down communism. In recent years, Americans of all backgrounds have rallied, marched, and protested to defend America from her original sin of racism—racism that has been transformed into a political phenomenon of trying to remake America into an autocracy.

To be sure, there are many heroes in this era of American history. Men and women of all races, backgrounds, and creeds have helped defend America in her darkest hour. History, however, will write that it was Black women—those underestimated sheroes—who stood center stage to lead America in pulling her back from the brink. We are the truth tellers. We are the fighters. To paraphrase the late, great Congressman John Lewis, we are the instigators of Good Trouble.

For four years, *our* voices held the Trump administration accountable in the rarefied, largely white, and complicit world of media and journalism, which often strained to construe the president's harmful words, behaviors, and decisions as "presidential" when his conduct was anything but—at great professional and personal cost.

Still, it was all in a day's work.

Black women have emerged as *the countervailing force* in our elections. Beyond the obvious and most recent example of Georgia, Black women in Selma, Alabama, propelled a record turnout in the 2017 special election that landed Democrat Doug Jones in the United States Senate (saving Democrats, *again*). Cornell Belcher, a political scientist and CEO of Brilliant Corners Research and Strategies, argued in an interview with me that the Selma race "foreshadowed the 2020 race [where we would see] a very different kind of turnout from African Americans, particularly African American women in the face of an existential threat to their communities and young people." According to exit polls, 98 percent of Black women and 93 percent of Black men voted for Jones. White women and men, by contrast, voted 34 percent and 26 percent for Jones, respectively. Black women's strong showing sparked commentary on social media and a celebration online, cascading to the point where a hashtag—#BlackWomen—trended on Twitter.

This book lifts up the experiences of these remarkable women, and so many others, to accelerate the arrival of the day when we hold equal power in our great democracy. Yes, *Black Women Will Save the World* is an ode to Black women. It is a portrait of our strength, our struggles, our triumphs, and our tragedies. While this book celebrates Black women, it does not take Black women for granted. *Black Women Will Save the World* is not just an overdue *Profiles in Courage* for Black women; it is an accounting of the costs that we incur as we serve our country, carry our families, and support our communities. My fervent hope is that we marshal the will and, indeed, *the allyship* necessary to change conditions for Black women across America by igniting a conversation about the many challenges Black women face individually, collectively, and *structurally*.

A cornerstone of this book is testimony from the country's and, indeed, the world's most powerful Black women. That focus is intentional. Many of the experiences of our sheroes are consistent with the challenges endured by Black women everywhere and every day. No matter where we sit, *all of us* battle the same slate of issues: we are ignored and overlooked in the world of work, denied opportunities for advancement, and denigrated and looked down upon in everyday interactions. The aggressions are both micro and macro. Our experiences speak to the "double whammy" that Shirley Chisholm spoke of as we shoulder an outsize share of the burden in our households and our workplaces—including on the front lines, where we are overrepresented in so-called essential work and are saving millions of lives. This book centers on the contradiction that defines the Black feminine experience: despite our education, our achievement, and our ambition, we are persistently and uniquely discounted in our pursuits, and too often we are made to feel as if we are not enough.

That feeling persists even though, empirically speaking, Black women are doing it all: at home, at work, in our community, and for our country. This book unfolds in three sections. The first section focuses on the experiences of Black women in leadership. For all our hardship and all our struggle, we live in a golden era of Black women in power. These chapters describe how Black women lead uniquely and powerfully.

The second section explains the costs of being Black *and* female. These chapters drive home the fact that Black women are perennially in the fight of an underdog. We are perpetually marginalized and discounted, if not counted out altogether. The truth is very few people even *listen* to Black women. This isn't my opinion, it's science.[5] Instead of being heard, Black women usually confront a range of micro- and macro-aggressions in everyday life, from work to social spaces. Our tone is policed, our appearances are judged, and our contributions are almost always minimized.[6] This is true even among the most powerful of us. Black women endure specific and frustrating treatment—what scholar and writer Moya Bailey describes as "misogynoir"—when both race *and* gender play roles in discrimination and bias. The price society exacts from Black women is very real and poorly understood.

Nevertheless, what is remarkable and worthy of elevation is that Black women have refused to internalize society's discriminatory treatment of us as a weakness. Black women subvert, challenge, and dismantle the very structures determined to undo us—*that* is our superpower and *that* is how we will save the world. When Black women fight, we don't fight for just us—we fight for everyone.

In our pursuit of a more equitable world, Black women bring all kinds of superpowers to the table. We are expert organizers; we always find a way to bring along our sisters, brothers, and communities. We are expert navigators; we code switch in order to navigate the tricky and often discriminatory dynamics of industry and politics. We are truth tellers; we mobilize the conscience of a country that is too slow to act on what's right.

And we pay attention. Cloaked in our invisible cape, we fly through the air to scan the horizon for injustice. We always find a way to save the day.

The third section focuses on The Future. The next generation of Black women already defies expectations, redefining what it means to be Black and female as they organize and defend our communities in new and powerful ways. As we look ahead to the decades remaining in this century, the power, influence, and agency of Black women will inevitably increase. For all these reasons and more, it's time that our larger society pays attention to what's coming in the next generation of Black women.

Indeed, a courageous conversation about the state of Black women is overdue. If 2020 taught us anything, it is this: the Black woman is undeniable. We have arrived (*hello, Madam Vice President*) and we are not going anywhere.

I recently interviewed the Smithsonian Institution Secretary Lonnie G. Bunch III, who said, "Black women have really been central to America's struggle for fairness. Kamala Harris is a symbol of that—the key will be how we build on her symbol."

I agree with Bunch. Black women have always been powerful. We have always been great. However, after our enduring centuries of oppression and decades of being overlooked, America is beginning to grasp the importance of the Black woman and reward her for her contributions to society. We are finally getting our due and the American public, really, for the first time, wants to understand what makes Black women so powerful—*and so great.*

I am so excited about The Conversation that *this* book will spur: a conversation about what it means to exalt and *respect* Black women; a conversation about what it means to celebrate our leadership and hold space for our vulnerability; a conversation about what true allyship really means; a conversation about honoring Black women by not just venerating us when we save the day (as we're apt to do) but rather by relieving us of oppression so that we can climb, advance, achieve, and triumph—so that we can breathe, heal, and *be.*

Black women have always "leaned in," and now you, my dear reader, must *lean in* to do your part: join me in *celebrating* Black women, *seeing* Black women, *supporting* Black women, and *standing* with Black women by creating space for their full humanity. America's superhero is here––she always has been. But now, we're paying attention.

Indeed, *Black women will save the world.* We already have. We are.

THE ASCENDANCY OF THE BLACK WOMAN (AND THE PRICE WE PAY)

When the Capitol insurrection unfolded on January 6, my colleagues and I reported the facts as they happened. I tracked the mob as it stormed the Capitol and as the mob desecrated the memorial for the late Congressman John Lewis. I learned and reported that the insurrectionists planned to hang the vice president. In horror, I reported that the Electoral College ballots had been whisked away to an undisclosed location just as the mob assumed control of the Capitol. For five hours, America and Democracy teetered on the edge of oblivion.

When I learned that explosive devices had also been discovered near the Capitol, I flashed back to 2018, when apparent explosive devices were mailed to former presidents Obama and Bill Clinton, as well as former Secretary of State Hillary Clinton and professional colleagues in the press. They were mailed by Cesar Sayoc, the man eventually convicted and sentenced to twenty years in prison for mailing the bombs. Earlier that year, I too had received an unmarked package from Sayoc.

My recollection of that day is vivid. I discarded the package when I first received it because I didn't recognize it. It had been forwarded to my home from my office in Pittsburgh by my then supervisor. He recognized the package when reports about the others who had received the packages made their way to news networks. Months later, the FBI warned me that I might receive an unmarked package from Sayoc. Indeed, the mysterious package I had discarded months before fit the description. It was a chilling moment. I was a target of an intimidation campaign, organized by Sayoc, to threaten and frighten President Trump's (perceived) political opponents and critics into submission. A coordinated mailing of potentially explosive devices to former presidents and journalists? What the hell . . . ?

That feeling was and is indescribable.

I received the package in the home where I raise my children. It arrived in the city that raised me, a community I love dearly. To put it bluntly, I received the package *where I felt safe*. That package and the countless death threats that led up to it—all incited directly or indirectly by the president—shattered that feeling of safety, perhaps forever. I haven't had the chance to tell the full story and chose to wait until I was whole enough to reveal it all. In late March 2020, Tennessee Congressman Steve Cohen invited me to testify before his committee about my experiences. That opportunity was postponed, most likely forever, because of COVID-19.

Here I share my truth.

Today I look over my shoulder in a way I never did before.

I have spent countless hours in therapy trying to reclaim my sense of peace and security.

Much of my hair fell out.

With the help of great doctors, therapists, and counselors, I've started to reclaim my sense of peace.

Most of my hair is back. My health, which deteriorated because of the stress, worry, and anxiety induced by the actions and deliberate choices of the Trump administration, has rebounded. Donald Trump literally drove me to drink.

Black women saved the world; Black women are saving the world. You better believe it. But we have paid an enormous price to do so. That's important to understand too.

There is a lot to unpack from the Trump presidency: his administration did untold harm to our environment; he neglected the most vulnerable in our society. More than any other president in American history, Trump abused power and flouted the law, while trying to destroy the very norms essential to the health and well-being of our democratic society.

But what is most striking and most personal to me is this: President Trump terrorized everyone who loves America, especially those who spoke up to defend America in her darkest hour.

Let that be a lesson: to save America is to serve America, *but for Black women, it is also to endure America.* In order to pull our country back from the brink, Black women have endured unspeakable hardship and horror.

And let me be clear: I love America. That is precisely why I insist on telling America the truth about herself. She deserves it because it's the only way she gets better.

Let us hope that with the benefit of hindsight and a long overdue celebration of the Black woman and her contributions to the country we love, that each of us improves. I hope that we—you—learn from this moment and grow: stop discounting Black women; listen to us, respect us.

If we do that, then maybe—just maybe—the next time America needs help, we won't have to work so hard to save her.

Section I

The
Power

HOW WE LEAD AND WHY

CHAPTER 1

The Superpower of Sisterhood

> *Being a Black woman who has the audacity to aspire to leadership, we're always going to be judged by a different set of standards.*
>
> Chicago Mayor Lori Lightfoot

SUMMER 2021: THE FIGHT FOR THE VOTE

Before I escaped to Martha's Vineyard to dip my feet in the chilly waters of the Inkwell, I was *doing the work,* as I have for decades, from Capitol Hill.

For more than thirty years, I have worked in journalism. Since the tender age of twenty-nine, I have served my country as a member the White House press corps. When I first started, Bill Clinton was president. Through the years, I've learned *a lot*—a lot about America, her people, her priorities, her angels, and her demons. While I dearly love and hold in high esteem the vast majority of my colleagues who serve alongside me in the press corps, my experience has been unique. Throughout much of my career, I have been the only Black woman in the press corps whose reporting focused on race and the many challenges facing Black Americans. More recently, outstanding colleagues like Yamiche Alcindor have joined me. But for a long time, I was the *only* Black woman who reported from the White House pressroom on the daily struggles of Black Americans.

Take a moment to think about that.

As the only Black woman reporting on the concerns and challenges of Black Americans, representation takes on a whole new meaning. I quickly learned to listen closely to the questions that are asked—*and not asked*. I learned to listen closely to the answers that are given—*and not given*. I learned to pay attention to the issues that are elevated—*and not* elevated. Most important, I learned early how to fight—first, for my seat at the table; and second, for the issues that are important to discuss at that table because they matter to the millions of people who are not in the room to ask the question, to challenge the answer, and to elevate the issue. *Those people* are counting on *me* to ensure that their struggle is represented in our great, messy, and often profoundly unequal democratic process.

My work is often inconvenient for people in power. As such, I've developed a thick skin. I've learned to compartmentalize the pain and hurt I feel when my contributions are dismissed or overlooked because of the color of my skin and my gender. I have learned to focus on what I can control because, ultimately, it's the only way to succeed in advancing The Work and the cause of what's right and just. And most important, I've learned to hustle—I never take a story, a source, or an opportunity for granted. The seat at the table—and the opportunity to fight for what's right—is never guaranteed. I have to earn it. Every day.

My career started in local radio in Baltimore, where I worked as a news announcer. From the beginning, I was good at my job and I *loved* it. I love using my voice to highlight the issues that are critically important to my community and our country. It felt good to participate in and not just passively listen to The Conversation. Despite my work ethic and early wins, it wasn't enough. I quickly encountered a glass ceiling on my career—a ceiling I was determined to shatter. I wanted to make more money in order to provide a better quality of life for my family. I wanted to have greater reach and impact in order to make a bigger difference for the American people. It wasn't long before I was hustling as a freelance journalist on the side and advocating for a promotion in my day job. I won on both accounts. I was promoted to news director and my freelancing led me to American Urban Radio Networks, where I was eventually named bureau chief and my career as a White House correspondent started.

Graduating from my local market in Baltimore to the national press corps in Washington, DC, was like moving from black and white to *technicolor*. It was a brave new world. The pace in Washington was faster and the stakes were so much higher. And the climate—and I am speaking candidly—was far more hostile. The racism and sexism I experienced at work was casual, omnipresent, and often destructive. Executing at an elite level in journalism depends on three things: indefatigable work ethic, terrific sources, and access to those in power. Unfortunately, in our society, while your gender and race have no bearing on your ability to work hard, they absolutely can interfere with your ability to cultivate sources and "get the interview" with the decision makers in power.

Unlike many in Washington who share the same pedigree vis-à-vis their private high school or college, I have far humbler roots. First and foremost, I am a Baltimore girl. I was born and raised in the city by my parents, Robert and Vivian Ryan, who were as proud as they were hardworking. They were working class: my father worked in trucking and my mother worked at Morgan State University for forty-two years in the student activities center. My parents instilled in me a work ethic and pride in family. I graduated from Morgan State University, one of Baltimore's two historically Black colleges, and a stint on the college radio station gave me my first journalism gig—those years lit the fire within me and forever changed my life.

The path I traveled to get where I am today was lonely, but I was never alone. I benefited enormously from the sisterhood and the role models of my sisters—my fellow Black women in the struggle and in the fight. I think about Isisara Bey, a producer and on-air radio personality, who gave me my start as journalist. I was a nontraditional bet: I came up outside of the National Association of Black Journalists unlike so many others who worked through the association, starting in college (or even high school). I didn't have the pedigree or the credentials that many use to validate their entry into the profession. I just had a hunger to tell stories about the truth in our communities and an intuition about my career that I learned to trust. I was born to report. I was born to broadcast. I've always known that this is my destiny.

I remember the first moment I knew I was going to be a journalist. It was 1976. Oprah Winfrey co-anchored the nightly news in Baltimore on WJZ TV. Watching Winfrey ignited a fire in me. She *looked* like me. She *sounded* like me. The way she reported the news mirrored how I imagined myself in my wildest dreams. She was— and is—my hero. I loved—and still love—her spirit. She's absolutely fearless. She asks the hard questions and always strikes at the heart of the matter. She's not afraid to incorporate her lived and personal experience in her reporting—she understands that her life's journey has value. For Black women, the personal is political. Her identity serves as her guidepost, and it directs her work.

I would not trade my particular life experiences, who I am, or the arduous road I have traveled for anything in the world. Like Winfrey, my race and my gender have had an undeniable effect on my work and career. I can say, candidly, that my journey has been harder than it has been for my white or male counterparts. On the other hand, because of my race and gender, my contributions as a journalist have been distinct and important. I learned long ago that *if I didn't ask certain questions, then they just wouldn't be asked.* Period. Questions about civil rights. Policing. Migration. Sex trafficking. Poverty. Fairness and equality before the law.

My focus on these issues has made me a target. This was especially true during the Trump years. In 2017, I was part of the reason why the social media hashtag #BlackWomenAtWork went viral after political commentator Bill O'Reilly and White House press secretary Sean Spicer dismissed, with all their condescension, Rep. Maxine Waters and me for being Black women effectively executing our jobs (for context, Spicer shut down my line of questioning during a White House press briefing because I was shaking my head during our exchange). Their attacks were personal, petty, and mean-spirited. I wish I could say that these incidents were unique or that they were a "shock to the system." They were not. The only difference is that the world had a conversation about our treatment and I would like to think that some Americans walked away from that conversation a little more enlightened about the level of discrimination that so many Black women still face just by showing up at work, every single day.

In the world of journalism, I am an endurance athlete. Through sheer force of will, I have stayed in my profession and climbed. My ascent hasn't always been welcome, but it has always been needed. In the summer of 2021, I was once again doing The Work by shining a light on people and priorities that much of the media would miss or overlook—to our collective detriment. In that July, the issue most on my mind was The Vote.

Our country is once again engaged in a generational struggle to secure our voting rights. There is no other issue as important—no challenge in foreign policy or even the COVID-19 pandemic is as serious and significant as The Vote. It's the currency of American Democracy. Ultimately, The Vote drives (or halts) all political progress in the United States. It's our country's bedrock.

That bedrock is cracking because American Democracy is imperiled. Trump and his allies at both the federal and state levels have made suppressing The Vote their dominant strategy for returning to power. While there are more of *us,* those that believe in a pluralistic, multi-racial democracy, than there are of *them*, those that would steer us toward autocracy defined by voter suppression and democratic exclusion of racial minorities, our advantage in representation is threatened if we aren't counted. Even though many Democrats let out a large sigh of relief after the November 2020 elections, Trump and his allies never stopped the assault on our Democracy. Trumpism continues to undermine The Vote as well as the policies and procedures that make fair and free elections possible in Congress, in state legislatures, and even in the courts.

For example, last year, in 2021, *just* between January and June, seventeen states enacted twenty-eight laws to restrict access to The Vote.[1] The pace of change at the state level exceeds America's most recent period of significant voter suppression in 2011, when fourteen states enacted nineteen laws to restrict The Vote (auspicious timing given the fact that our first Black president was seeking reelection in 2012). By the end of 2021, *nineteen* states enacted *thirty four* laws to restrict access to The Vote. A record that is likely to be broken in 2022 as Republicans lay the groundwork to "Stop the Steal," again, in 2024.[2]

One hundred leading scholars released a statement warning that Republican-led states were proposing or implementing "radical changes" to election laws that risk transforming our political system into one that "no longer meets the minimum conditions for free and fair elections." Their statement was ominous: "our entire democracy is now at risk."[3]

Let's confront the facts. Nearly all these bills and laws flow from one source: the Big Lie. When you study the record and the public comments that surround these measures, you learn the following: politicians and so-called activists parrot the former president's accusations, and Trump and his allies are determined to prevent a repeat of the historic turnout in 2020, organized by Black women throughout the country, that propelled Joe Biden and Kamala Harris to the White House.

On Capitol Hill, Black women were the first to sound the alarm as Trump and his allies tried another assault on American Democracy. As Black women, we are forever vigilant. It was obvious that following his loss, then President Trump would stop at nothing to hold onto power. First, he tried to interfere with the electoral count in Georgia, Arizona, and other states where he had expected to win (but didn't). Of course, there was the white nationalist insurrection on January 6 to "Stop the Steal."

In the months and weeks leading up to January 6, as large elements of America's political system rationalized Trump's increasingly dangerous behavior, "Auntie Maxine" Waters spoke plainly to me about the existential threats facing the nation:

> **I've always believed that the president of the United States wants to create confrontation and chaos. I've always believed that he's capable of initiating and encouraging civil war. I don't know how many [are coming to the Capitol] or how big it's going to be, but I believe it's potentially dangerous. The white supremacists, the KKK, the Proud Boys, the Oath Keepers—they are accustomed to carrying arms.... And I believe some of them will have arms (on January 6). Whatever happens, and whatever violence takes place, it is squarely the responsibility of the president.**

Well before January 6, Representative Waters understood what most Americans have only recently come to understand: America—and its always-fragile democracy—requires ever-vigilant defense.

Waters even notified security at the Capitol that an attack on the Capitol and our electoral process was imminent. Other Black women in Congress, including Massachusetts Rep. Ayanna Pressley, used their voices to call attention to the nefarious work of Republican-led state legislatures that immediately set out to ensure a Republican capture of the White House in 2024.

These Black women are using the tools the founders left us. The Constitution articulates ways to disqualify a president on account of their conduct. The founders appreciated, far better than *some* of us do today, that democracies are susceptible—even existentially vulnerable—to the forces of populism and demagoguery. To quote Winston Churchill:

"Democracy is the worst form of government except for all those other forms that have been tried." Lest we forget, Adolf Hitler rose to power first through democratic processes and popular elections.

In this light, that is why truth tellers, like representatives Waters and Pressley, are so vitally important. And it's powerfully important to recognize the role Black women play as the guardians of our American democracy as those truth tellers. When Waters called for Trump's impeachment (the first time), she did so to save America from herself and not to play politics as it was often cast by the media and political pundits. I saw more of my colleagues come around to the perspective, especially during the second impeachment proceeding, that Trump posed an existential threat to American democracy. That's encouraging. However, it's important to acknowledge the extent to which Black women and their voices were discounted during the Trump administration. Stepping back, we should appreciate how Black women experience marginalization *and* how society is impacted as a result of our exclusion. Let's not forget that we speak up *to make a difference*. We aren't doing this for our health, y'all. Yet, when we are ignored while trying to speak up, the experience is akin to gaslighting. As for society? Well, when we're excluded or marginalized the results are usually devastating.

JULY 2021: PROTEST ON CAPITOL HILL

Without a doubt, The Vote is our chief battleground in the fight for American Democracy. Even our adversaries overseas understand this. Namely, Russia, China, and North Korea make it their business to meddle in our elections and undermine our confidence in our institutions. Ironically, Trump can demagogue and polarize his supporters to undermine The Vote and the American Constitution *precisely because* we live in a free country. For all these reasons and more, we must fight back as citizens and leaders committed to upholding our country's foundational principle of free and fair elections.

If there is one through line in my reporting, it is the commitment to defend America's voting rights. I am a child of the 1960s, raised by Black parents in my native Baltimore. I understand from firsthand experience how important it is that we have the right to participate in our democracy—freely and fairly. Our ancestors fought to secure the right; our foremothers and forefathers fought to defend it. It is so today: we must protect our place in this country as equal citizens.

Our country works because we work to protect her. And without vigilance, we run the risk of losing all that makes America truly great.

Black women understand this because we've always had to. From the American Revolution to the Civil War to civil rights, Black women have always played the role of defenders of freedom, fairness, and justice—the hallmarks of our democracy—from its archenemies: fear, terror, division, and discrimination. As we learned (again) during the Trump administration, these are the tools of autocrats. We have to battle their hate with the tools pioneered by our ancestors. The Rev. Dr. Martin Luther King Jr. reminded us decades ago, "Darkness cannot drive out darkness; only light can do that. Hate cannot drive out hate; only love can do that."

Black women have lived out these words; indeed, they are our mantra throughout the entirety of American history. Here, in 2021, Black women and the power of their example are driving out the darkness that continues to threaten The Vote and our democracy.

And while (white) Democratic senators dithered as to whether to abolish the filibuster in order to secure The Vote for all Americans,

Black women put their bodies on the line to highlight the threat and demand action.

In July 2021, ten Black women, many of them members of Congress, peacefully protested the Capitol's inaction on voting rights. For their service, as equal members of America's esteemed and deliberative body, they were arrested at the Capitol for demonstrating. Congressional Black Caucus Chair Rep. Joyce Beatty of Ohio and Melanie Campbell of the Black Women's Roundtable led the women in that protest. Like the great civil rights activists before them, those ten Black women, many of them older in their years, were detained and led to jail. Campbell channeled the civil rights heroes of old in an interview with me about the July demonstration:

> For the Black vote, we're the ones who mobilize because we don't just go on our own, we take our families, our friends, and communities with us. We lead, not just vote. We lead. When we talk about the impact of what will happen with our voter rights being really eroded in such a massive way, when you're talking ability or possibility of forty-eight states and not having a federal intervention, then we're talking about the erosion of Black political power. So, that's part of the sense of urgency and we had to, and have to. Not just this week, but we're going to stay and sustain it. We're going to support.
>
> I'm ready to go again. I'll go again and again until we get our voter rights. Back in my spirit, I'm there. I'm there in a way that I've not had to be there. I've supported protests, in my other years, early years, I was active in organizing protests and going to jail, but my organization has not been known as a direct action, so I try to play my position as far as organizational. But I could not do this; it was just in my spirit. And it still is. I'm ready, let's do this. Let's keep

pushing, not that it's the only solution, but yeah, go to the White House, talk to the administration. Let's go over here and push these senators. Let's get everything we need to have. Keep the faith.

Change happens. Good trouble. As we think about John Lewis. Black women have always been, what do they call us? The backbone, I don't like that term. Not just political, but of the movement. What's different about the power of Black women is that we're not just behind the scenes doing, we are leading, and we are unapologetic about that, that's why we demanded and we challenged this new administration. If you want to win, you need a Black woman running mate, and our current president was a smart man and he did it and look here, we have the first Black South Asian woman vice president. So we know when we make up our minds as Black women, we can make it happen. We are unstoppable. We've had to be. It's in our DNA. I think that's the part. It's not like those who came before, they fought the fight the way they had to fight in their time. In our time, we have to be bolder about it, it's okay that I'm aggressive.

We're going to win, April. We're going to win; we have to win. The future of our nation is at stake and Black women have the ability, by pulling folks together, to help save—once again—help save our nation from peril.

President Biden rightly equated today's fight for voting rights with the reason for the Civil War (and lest we forget that Harriet Tubman, an officer in the Union Army, led troops into combat during the Civil War—Black women will do *that* too, if it comes to it). The president makes an important connection and one that all Americans should pay more attention to: this is a generational struggle, one that repeats and seemingly never ends. *Backlash* is a defining feature of the United

States. In America, progress for all people, especially Black and brown people, is never permanent. It is always won. It is always earned.

The defeat of Trump and Trumpism in 2020 did not mark an end—it was only the beginning in what is sure to be a long fight for what's right. Trumpism isn't going anywhere. Therefore, Campbell teaches us that vigilance is our watchword.

Their brave demonstration was peaceful. I sent a reporter to cover the event. Unlike the destructive and chaotic events of January 6 (when an angry mob of thousands of people, most of whom were white, stormed the Capitol with impunity), these ten women chanted and sang in a beautiful act of civil disobedience that started in the Hart Senate Office Building. And it almost didn't happen. Without a few key members of the press corps to highlight their agenda (namely, a handful of reporters of color and women of color), these women would not have even been allowed to enter the Senate building and it is likely that their work would have been ignored.

Those brave women responded to President Biden's exhortation that the fight for voting rights was critically important but "needed more urgency" from Congress. This call to action was reminiscent of President Lyndon Johnson's demand to Dr. King during the battle for civil rights: "go and make me do it."

So they did.

Black women from across the generations joined Rep. Beatty and Campbell, including Tamika Mallory, Cora Masters Barry, LaTosha Brown, Barbara Arnwine, and Barbara Williams-Skinner—all leading advocates for civil rights. Some of the women were detained with their arms behind their backs and their wrists tied together tightly with painful zip ties, and then they were loaded into police vans with their arms painfully contorted.

For some of the women, it was not the first time they had been arrested. Once in jail, Beatty spoke to the women about what was next. With uplifting stories and songs, she distracted the younger, less experienced women from focusing on the miserable conditions in which they found themselves. Many of them had to use the bathroom, only to stare at the awful open toilet in the center of the cell. Others lay down on a steel bed without pillows or sheets, their heads aching from relentless migraines.

Beatty thought of Dr. King and his historic "Letter from a Birmingham Jail" in which he addressed his critics in the clergy:

> I wish you had commended the Negro sit-in-ers and demonstrators of Birmingham for their sublime courage, their willingness to suffer and their amazing discipline in the midst of great provocation. One day the South will recognize its real heroes. They will be the James Merediths, with the noble sense of purpose that enables them to face jeering and hostile mobs, and with the agonizing loneliness that characterizes the life of the pioneer. They will be old, oppressed, battered Negro women, symbolized in a seventy-two-year-old woman in Montgomery, Alabama, who rose up with a sense of dignity and with her people decided not to ride segregated buses, and who responded with ungrammatical profundity to one who inquired about her weariness: "My feets is tired, but my soul is at rest." They will be the young high school and college students, the young ministers of the gospel and a host of their elders, courageously and nonviolently sitting in at lunch counters and willingly going to jail for conscience' sake. One day the South will know that when these disinherited children of God sat down at lunch counters, they were in reality standing up for what is best in the American dream and

for the most sacred values in our Judeo-Christian heritage, thereby bringing our nation back to those great wells of democracy which were dug deep by the founding fathers in their formulation of the Constitution and the Declaration of Independence.[4]

Beatty's connection to Dr. King is important because the audience for that letter weren't his "enemies"—the segregationists or members of the opposition. Rather, Dr. King was addressing his "friends," the members of the clergy who refused to act with urgency when it was all on the line. Beatty and her compatriots were not taking a stand to challenge former President Trump or his supporters. No, the congresswoman and the nine other women who joined her were rousing the consciousness of Democrats—especially those who refuse to act. Joe Manchin. Kyrsten Sinema.

The jailed women were eventually released on bail. Immediately after, I spoke to some of them. We were all struck by the juxtaposition of their treatment as peaceful demonstrators for voting rights and the mayhem that unfolded on January 6 without consequence or accountability for the participants. The primary difference, of course, is that the individuals arrested in July were Black women. Their cause? Our civil rights.

It doesn't make any damn sense.

Another difference? The Black women who risked their bodies for civil rights in July had superpowers available to them, namely, *sisterhood*. That "powerful act of solidarity," Tamika Mallory referenced after her colleagues were arrested, echoes Rep. Ayanna Pressley's sentiments about "the Squad," a powerful group of women of color who are redefining political leadership right before our eyes. "Our squad is big. Our squad includes any person committed to building a more equitable and just world . . . and given the size of this squad and this great nation, we cannot, we will not be silenced."

In her piece "Sisterhood: Political Solidarity between Women," bell hooks writes poetically about the power of sisterhood, challenging the notion that we cannot come together to fight our common enemies:

> **Solidarity strengthens resistance struggle. There can be no mass-based feminist movement to end sexist oppression without a united front—women must take the initiative and demonstrate the power of solidarity. Unless we can show that barriers separating women can be eliminated, that solidarity can exist, we cannot hope to change and transform society as a whole.[5]**

The late, great sister hooks warned us about the elements of white supremacy that have held back the cause of feminism and sisterhood, pointing out that many white women participate in the vicious oppression of Black women at work and at home. She also called on us to recognize the unique oppression of Black women—not to isolate Black women from the broader feminist movement but to empower Black women as the leaders of a righteous movement dedicated to the freedom and dignity of all people.

In an interview with me, Chicago Mayor Lori Lightfoot channeled bell hooks by speaking to the power of sisterhood in driving political change on the front lines of American cities. "We call each other our *sister mayors*, and that's what we feel," said Lightfoot. "We are really strong and united. We know exactly what's up, the challenges that we're facing. And we are incredibly supportive of each other." Lightfoot spoke at length about the unique pressures on the Black women across the country who were in the mayor's seat. She highlighted the double standards and uniquely unfair treatment Black women endure in our current battle against Hate.

She continued:

> Being a Black woman who has the audacity to aspire to leadership, we're always going to be judged by a different set of standards. And the lens on us is going to be different [from] the lens on white men who have occupied the same roles. We can't let that be the choice. My parents, and my mother in particular, taught me that from a very young age, when you're a Black woman, the cavalry's not coming to save you. You've got to be very clear about who you are, what your values are, and you've got to forge ahead. And you cannot let the critics destroy you because they're always going to be there and they're loud. You mentioned Black Americans across the country; it's not a coincidence to me that the former president attacked us: me, Muriel Bowser, Keisha Lance Bottoms, other Black women mayors across the country, a spokesperson using our names every single day in press conferences. You were there. And what that does, when the dog whistle was blown by Trump and his people, people came out of the woodwork to attack us. But what that has done is make our bonds even stronger from coast to coast. From London Breed to Muriel Bowser to Keisha Lance Bottoms to LaToya Cantrell.

Sisterhood is essential to the salvation of Black women because the experience of being a Black woman is uniquely challenging. The forces of racism and sexism demand that Black women endure a particular *pressure* that requires unbreakable strength, individually and collectively. Much like a diamond held together by unbreakable bonds, Black women must hold together in order to fully realize their freedom. As LaTosha Brown, cofounder of Black Voters Matter, says:

> I think of Black women as diamonds. A diamond is a piece of coal under extreme pressure. The value of a diamond is Clarity. And Hardness. For us, as Black women, the pressure that we have felt is the traumatic circumstance of racism and sexism in this country. Pressure can do one of two things: it can crush and destroy you, or it can transform you. Black women have been able to literally take this pressure and allow it to transform ourselves.
>
> There is something about sisterhood. When we work in our sisterhood, it creates that hardness. Sometimes when I am unsure about myself, there's always some sister to come along and say, "You look good!" or something as simple as "I like that lipstick on you" or "Preach, Mama." Sometimes all I need is a word from a Black woman to feel affirmed. Sometimes all I need is to hear a sister's voice in a room.

Black women have been standing strong, together, for more than four hundred years in this country. Sisterhood is a defining quality of our leadership and as recent years have shown us, our leadership is rare and needed. Courage is in short supply.

THE STATE OF BLACK WOMEN IN POLITICAL LEADERSHIP

In 1968, Shirley Chisholm declared and won her race for Congress, the first Black woman elected to walk the halls of Congress. Today,

Black women represent nearly 5 percent of the United States Congress, with twenty-five Black women serving in the US House of Representatives. With Vice President Kamala Harris's ascension to the executive branch, there are no Black women in the US Senate. There are also no Black women represented in the Republican Party on Capitol Hill. The twenty-five members in office represent more than half of all Black women ever elected to Congress (forty-seven Black women have been elected to Congress since Chisholm's pathbreaking election).

The Black women in power today hold disproportionate sway and influence. Rep. Maxine Waters is one of America's Truth Tellers. Rep. Ayanna Pressley co-leads the Squad, channeling the voice of progressives from across the land. Rep. Beatty, the chair of the Congressional Black Caucus, guides the caucus that single-handedly delivered the presidential nomination to then candidate Joe Biden.

I wish I could deliver better news from other spheres of American life. Take corporate America. Only three Black women have ever served as CEO of a Fortune 500 company. And it's not for lack of trying. Between 1977 and 2015, 532 Black women have graduated from Harvard Business School, one of America's premier hubs of future leaders. Sixty-seven of those women attained some rank in the C-Suite, including the title of CEO or chair. When these leaders are interviewed and their careers are studied, a few themes emerge: Black women, by virtue of being different because of their race and gender, battle a pandemic of "pattern matching" that locks them out of opportunity. Black women and their immense talents are often overlooked or ignored altogether because they don't look like the leaders who have come before them. When Black women succeed against the odds, their resilience[6]—that is, their emotional intelligence, authenticity, and agility—sets them apart.

As the *Harvard Business Review* found, the Black women who excel in leadership are experts at emotional intelligence, managing against "identity abrasions" (the negative interactions Black women endure because of who they are) that threaten to undermine their confidence and sense of competence. As one explained, "We were all told that you had to be smarter or run faster or jump higher or be better than anybody else around you just to stay in the game. That was a lesson from early, early on—from my parents, teachers, mentors, church. So, you come [to your job] with that orientation."[7]

Black women in leadership use sisterhood to confront the "visibility/invisibility conundrum," the phenomenon of being both highly visible and completely invisible because of their identity. This is isolating. Women who are effective at breaking through the conundrum do so by investing in sisterhood *and* brotherhood, transcending differences in identity to provide a way for others. It's in our DNA. More than any other demographic, Black women go out of their way to "provide opportunities to their subordinates to produce professional growth."[8]

In a large study about the leadership of Black women, we overwhelmingly characterized our own leadership style as "ethical," "respectful," and, finally, "nurturing." As Black women, a defining trait of our leadership is to make it a little easier for others and those coming up behind us. Simply, we pay it forward. We look out for those on the "come up," no matter their gender or race, ensuring that those with less power have the opportunity to get ahead. When pressed, Black women express feeling marginalized and overlooked in their own careers, and reject perpetuating that cycle for others.[9]

It's really that simple.

It's the Golden Rule. We treat others the way we wish to be treated. Imagine where we would be if *everyone* did that, right?

In addition to investing in a support system and cultivating the next generation, Black women also create space for challenging, authentic conversations about our struggle and *the* struggle. We elevate our voice and use it to make an impact. As LaTosha Brown said:

> **Our voice is how we carry and protect and nurture our humanity. That was the only thing. Even with our language and identity being stripped, our bodies being used as just a utility for getting work done . . . the one thing they could not control is our voice. For me, this work is an extension of not just what I believe around democracy or electoral work but . . . for Black people to assert our humanity. How do I lend my voice and create space for our humanity to be recognized and to flourish? Democracy is not the end goal for me. The end goal for me is for our people to have the freedom of movement, to have the freedom to radically reimagine everything around us, to be able to freely express and stand in our humanity.**

The ability to authentically communicate our humanity and its connection to what Dr. King called the "inescapable network of mutuality, tied in a single garment of destiny" is essential to the successful leadership of Black women. Our genius is to tie our particular struggle to the universal. *That* is The Work of leadership and why just a handful of Black women in the United States Congress wield such outsize influence. Our voice is our greatest gift. It empowers us to punch above our weight and, just as important, to continue to forge a path forward for everyone. When one of us wins, we all win.

To Brown's earlier point about the Black "diamond," Black women routinely transform a crisis of confidence into the fuel for radical self-determination. More than any other demographic, Black women grapple with feelings of both self-doubt *and* self-determination. These feelings are interrelated, fueling one another. As the world conveys to Black women that we should remain silent or meek or marginalized or ignored, we process those feelings in order to project confidence and a sense of our own inevitability. As the literature says, we are "self-driven" in large part because we actively grappled with the doubts and lack of confidence society engenders in us. This ability to transform pressure or negativity into positivity and fuel is yet another unique leadership trait that Black women tap into for the benefit of everyone.

Black women have always been more effective because of their collective orientation. This is especially the case in the fight for voting rights. Here's an excerpt from "Between Two Worlds: Black Women and the Fight for Voting Rights":

> During the 19th and 20th centuries, Black women played an active role in the struggle for universal suffrage. They participated in political meetings and organized political societies. African American women attended political conventions at their local churches where they planned strategies to gain the right to vote. In the late 1800s, more Black women worked for churches, newspapers, secondary schools, and colleges, which gave them a larger platform to promote their ideas . . . Because of their unique position, Black women tended to focus on human rights and universal suffrage, rather than suffrage solely for African Americans or for women.[10]

Black suffragists, like Mary Ann Shadd Cary, both fought for and were critical of the Fifteenth Amendment—which established the right to vote (albeit only for men)—because it *did* represent a huge step forward for Black men *and*, nevertheless, excluded women.

Another Black suffragist, Nannie Helen Burroughs, wrote and spoke extensively about the importance of women cooperating, across lines of race and class to achieve the right to vote for all women. Even then, Black women of the era practiced a sophisticated calculus of fighting for the greatest good possible in that political moment—the political enfranchisement of Black men—while advocating for the cause for all women. That ability to fight for the greater good while sacrificing for the immediate win is a hallmark of Black female leadership and it was indispensable to the social justice victories of the nineteenth century. Without the activism and dedication of Black women, it is unlikely that those historic amendments to end slavery, establish citizenship rights as well as guarantee equal protection under the law, and extend the right to vote would have been enshrined in the Constitution.

That pattern continued throughout the twentieth century. Here's another excerpt, this time from "Stacey Abrams and Other Georgia Organizers Are Part of a Long—But Often Overlooked—Tradition of Black Women Working for the Vote":

> In the 1920s and 1930s, Mary Church Terrell and Nannie Helen Burroughs were key leaders . . . for Black women voters . . . Other Black women created their own newspapers, magazines, and pamphlets, to publicize their fight for voting rights.
>
> Black women carried on these voter-education efforts through the early 1960s, at churches and bus stops and beauty shops, on farms and at community meetings. But they rarely became household names . . . male leaders of the fight for voting rights were the spokespeople who talked to and got quoted in news outlets . . . there was a lot of chauvinism, as the late Congressman John Lewis pointed out in his memoir.[11]

One of the most powerful leaders for Black women's voting rights emerged during the civil rights era when a sharecropper in her forties, Fannie Lou Hamer, discovered at a community meeting in a church that she could register to vote. Although Hamer lost her job for doing so, she became one of the most important activists of the 1960s. She served as the field secretary for the Student Nonviolent Coordinating Committee (SNCC), traveling the country to educate other Black farmworkers about their rights.

"When Hamer became aware of her constitutional rights, she was determined to use them," historian Keisha N. Blain, who is working on a biography of Hamer, shared in *Time* magazine.[12] "But even more," Blain goes on, "she wanted to ensure that others would also benefit from this knowledge."[13]

Super Vote-Getter Stacey Abrams has followed in Hamer's footsteps, continuing a tradition that dates back to the Black suffragists of the nineteenth century. Abrams is on the front lines of fighting against the effects of decisions such as the 2013 Supreme Court ruling that invalidated a key provision of the 1965 Voting Rights Act. Abrams's efforts culminated in 2020, when Georgia voters, fueled by the historic turnout of Black women, delivered Democrats Rev. Raphael Warnock and Jon Ossoff to the United States Senate (thereby delivering a Congressional majority to the new Democratic president). Her organization, Fair Fight, raised an astonishing $6 million in *just three days* for those historic Senate races in Georgia.

Here's another excerpt from "'Whatever It Takes': How Black Women Fought to Mobilize America's Voters":

> Democrats have long pointed to Black voters, more specifically, Black women, as a crucial voting bloc, decisive to elections since

former president Bill Clinton's victories in the 1990s. But this November, successfully flipping the southern, Republican-led state of Georgia to the Democrats for the first time in 28 years has drawn attention to the organizational power of Black women, whose large-scale mobilization efforts appear to have resulted in massive turnout among people of color in those cities, experts say.[14]

The importance of that work cannot be overstated. The underappreciated work of Black women to get out The Vote has most recently empowered two years of groundbreaking legislation from Capitol Hill. History has shown us that when Black women fight to open the door for themselves, they are fighting to open the door for *everyone*—ultimately driving progress for all people and communities. It's a tradition as old as this country. It's time we recognize it as such.

CHAPTER 2

How Not to Be Erased

> *You're going to watch me work.*
> Alicia Garza, cofounder of the international Black Lives Matter movement

FALL 2021: REPORTING ON MIGRATION

Erasure is something all Black women know something about. Never in my life have I encountered a Black woman who hasn't experienced a micro- or macro-aggression that tries to marginalize us in conversation or force us out of the picture altogether. In day-to-day interactions, at the doctor's office, or at work, we're always fighting for our place—we're always fighting to be seen, heard, and respected.

The *Oxford English Dictionary* defines erasure as the "removal of all traces of something; obliteration." That word *obliteration* gets me every time. For the four hundred years, we have been in America and so many forces have tried to obliterate us. Why?

Throughout much of her history, America has understood large aspects of its identity through the prism of race and gender, with *racism* and *sexism* animating many of the country's popular narratives. With that as context, there is no greater threat to the systems of patriarchy and white supremacy than a successful Black woman. Indeed, when they are successful, then how can discriminatory beliefs endure? When they flourish, the strengths and talents of Black women contradict and force the unraveling of hateful belief systems.

I think that's why so many Black women in leadership moved quickly to fight the Big Lie. We *know* white supremacy and the patriarchy when we see them, and we are driven to oppose them wherever they manifest, whatever shape they take. Nikole Hannah-Jones, most famously known for *The 1619 Project*, her wonderful examination of the legacy of slavery in the US that appeared in the *New York Times*, said to *CBS This Morning* that "all journalists are activists."

For Black journalists, especially Black women, her words are gospel. Our activism is rooted in serving the underserved, the forgotten, the marginalized. I used to push against the description of my work as "activism" because that adjective was lobbied to delegitimize my work. But in time, I have come to accept and embrace the description. Hannah-Jones is right—all journalism is activism and what matters most is the cause for which you are an activist. Ultimately, for whom do you fight?

A poignantly detailed fact that is little known is that Black journalists were, at one point, encouraged by the federal government to serve in the highest spaces and places to ask the questions that reveal the truths, hurts, and oppression faced daily by the communities not seen or adequately cared for by the broader society. In 1968, a year after I was born, then President Johnson called for a commission to study unrest in some American cities. That unrest was due to inflamed tensions between the police and the Black community. This sounds all too familiar, a cycle of violence that has not been fixed in hundreds of years. The report published by the Kerner Commission called for including more Black journalists in the White House press corps to help remedy the gap between the powerful and the (too) many communities struggling under the yoke of oppression.

But instead of seeing more Black journalists welcomed into the space, too many of us have battled erasure. Indeed, as the only Black woman covering Black issues in the White House press corps until 2018, I have confronted this notion of erasure throughout my career. I learned early that if I didn't find a way to make space for myself in The Conversation, then I was effectively erased from The Conversation altogether. Politicians, policy makers, hell, even other journalists were not going to make space for me "at the table." I am good at making space for myself, whether you like it or not.

Alicia Garza channels this spirit:

> **Eight years ago, when we started the Black Lives Matter Global Network, people thought that we were crazy and people thought that we were being too radical. It was like, why do we have to say Black Lives Matter? Why can't we say Black Lives Matter too? And now, eight years later, we have transformed the globe. And so, I think that one of the contributions that Black women make is the audacity to say, "You're not going to tell me that I can't do it. You're going to watch me work."**

I agree with Garza. We take up space for ourselves, for our communities, for what's right. As Garza says:

We do it for ourselves. Black women sit at the intersections of so many of the rigged rules that organize our lives, whether it be our access to protections in the workplace, whether it be access to protections in the home, whether it be access to health care, access to childcare. Black women are at the losing end of so many of the disparities that rigged rules in this country create. And so, when Black women motivate and activate and mobilize to save democracy, to bring democracy to bear for the first time, it's because we need it in our lives. But we also do it for our communities and we do it for the people that we love.

As I embark on my twenty-fifth year in the White House press corps, one of my "favorite" pastimes is combating the erasure of Black women. I get plenty of practice because I deal with it nearly every day at work. I almost get a kick out of it because the same basic pattern has unfolded too many times for me to count. I have learned to laugh at it, correct the record, and move on. I'll give you an example.

I asked Jen Psaki, President Biden's first press secretary, a question about the US deportation of Haitian migrants in the wake of the natural and political disasters that have forced countless families from their homes and out of the country to seek asylum:

Whether it's the Haitians that left in 2010 for South America and are now trying to travel here, or the ones who fled after this earthquake, *what is there for them to go back to?* The nation is in unrest. The president was assassinated. The people are scared of gangs and Democratic rule is not necessarily in place. Elections will not be held on time. And then there is the issue of the earthquake. *What are you deporting them back to?*

I asked the question to highlight the pain of the Haitian people and to draw attention to the issue of mass migration—South Americans, Haitians, Africans, people from all over the world who have been fleeing horrific situations in their home countries in pursuit of their very survival. I felt it was important to highlight the disparate treatment of Black asylum seekers, the Haitians, relative to the treatment of other communities. Our great country has had many admirable moments of embracing refugees from across the world, from Ireland to Eastern Europe, and, most recently, from Afghanistan. But when the asylum seeker is Black, the treatment, historically and today, is radically different. That is important to name.

For this reason, I wanted to hold the Biden administration accountable. The images of the immigration enforcement's "horse patrol" striking Haitian refugees with either whips or long reins are despicable. Sunny Hostin, the intrepid cohost on *The View*, whose husband is Haitian, is working to humanize the Haitian refugees who are so easily dismissed by the mainstream media and throughout American culture. Hostin vacations in Haiti with her family; she has reverence for the beautiful island and its unique culture. Through her familial bonds and as a student of history, she also appreciates how we, as Americans, have treated Haitians throughout history. Over the years, we've dehumanized Black migrants: far from "give me your tired, your poor, your huddled masses yearning to breathe free," America's treatment of Black migrants echoes the treatment of Black people through the ages extending to and including the treatment of enslaved Africans in America.

This context is critical and guides reporting that helps shape The Conversation. Throughout my career, I have made these kinds of questions central to my reporting. Again, if I do not ask *these* questions, then there's a very good chance that these questions don't get asked. I try to offer context about why the question is important and ask the question that strikes at the heart of the matter. In the case of the Haitians, what is there to deport them to? Where is the legendary compassion that America, at her best, is known for throughout the world?

The good news is that the question went viral, gaining traction on social media, and was picked up by several major news outlets. This helped lead to some change and some acknowledgment by the Biden administration. The federal Horse Patrol Unit officers were placed on administrative leave and the horse patrols ended. Not enough, but a start.

The bad news is that I was erased from the broader coverage. Across news outlets, "a reporter asked Jen Psaki about the reality for Haitian migrants." *This* always happens. Unlike my great colleagues and personal friends who are elevated in the coverage for making important and incisive points, I am not. My identity as a Black woman is always erased from The Conversation. Even when we ignite the discussion. Let me be blunt: many people simply do not want to give Black women credit. This is especially true if you do not share the pedigree of the elite—no Harvard degree here. I am a proud and vocal graduate of Morgan State University, an extraordinary and excellent historically Black college.

THE DISAPPEARING BLACK WOMAN

When you're charting a path that no one else is charting, people tend to be dismissive. When I first started reporting, I was instantly branded as "militant" for asking questions that centered humanity as I interrogated the powerful. I guess I was supposed to "stay in my place," but that "place" wasn't for me. As I advanced in my career, I learned to take up space and insist on my recognition, the credit I deserved.

I did not, and do not, do this because of a sense of vanity. No, I take up space because this is How Not to Be Erased. I insist on claiming my space precisely because so many elements of society prefer me to disappear altogether. In the Trump era, with hate and fascism on the rise, taking up space in The Conversation, being loud about my point of view, and insisting on being heard *saved my life*. The Trump era was the era of negation for Black women; we were supposed to, simply, disappear.

Lest you think this is me being sensitive, the erasure of Black women is a broadly chronicled empirical and historical phenomenon. Black women have made enormous contributions to society and yet, reliably, we are excluded from media coverage, the public record, and the history books. We are even erased in the mind. The mental shortcuts of implicit bias lead many, even the well-meaning, to discount or ignore our presence or our voices altogether.

Erasure is especially common across the worlds of activism and social justice movements. When you examine the feminist movement, many of the struggles and the gains that followed for women were instigated by Black women and yet rarely are our contributions acknowledged, let alone celebrated.

For example, the origin of antirape activism dates back to 1866, when a group of Black women testified to the US Congress about the experience of being gang raped by white men during the Memphis riot.[1] Despite the women's testimony, the perpetrators were never brought to justice. The bravery of those women led Fannie Barrier Williams and Ida B. Wells to found and drive campaigns dedicated to ending sexual violence against Black women and girls. Their work laid the foundation for the broader feminist movement in the 1970s that led to many reforms, including the creation of rape crisis centers and college campus activism around the notion of consent.[2]

Yet the contribution of the pioneering Black women has been erased from our recollection of how these important changes came to pass, with white women occupying most of the space.

This harmonizes with the experience of Tarana Burke, the founder of the #MeToo movement. A Black woman, Burke quickly saw the movement she founded center the experiences and narratives of prominent white women.

Here's an excerpt from "Black Women Often Ignored by Social Justice Movements":

> "Black women are often overlooked in people's conversations about racism and sexism even though they face a unique combination of both of these forms of discrimination simultaneously," said lead researcher Stewart Coles, a PhD candidate at the University of Michigan's Department of Communication and Media. "This 'intersectional invisibility' means that movements that are supposed to help Black women may be contributing to their marginalization."
>
> Feminist movements that focus only on issues that predominantly affect White women without addressing racialized sexism ignore the needs of Black women, who face higher rates of police abuses, including sexual violence, Coles said.[3]

Coles concludes that the key is to listen to Black women about their concerns and center their experience in any movement for social justice.

Unfortunately, we still have a lot of work to do if we're going to elevate the pain and the promise of Black women as a central concern in our collective fight for social justice. Another movement—#defundthepolice—is following a similar pattern. In 2014, Black women came together to #SayHerName to elevate the fact that Black women are routinely beaten, killed, raped, or assaulted by the police, and yet, our stories attract a fraction of the attention. Following the tragic murder of George Floyd, #SayHerName forced the public to remember the tragic murder of Breonna Taylor whose death, initially, came and went without a murmur of public outcry.

Here's an excerpt from "Fighting 'Erasure.'"

> In December, Daniel Holtzclaw, a former Oklahoma City police officer, stood trial for sexually assaulting 12 Black women and one teenager. He preyed on the vulnerable—the poor or drug-addicted or those with outstanding warrants—threatening them with arrest if they wouldn't comply. Few people were following the case, however, until Black women on social media began calling out the press for ignoring the story . . . "ERASURE IS VIOLENCE." Deborah Douglas, writing for *Ebony* magazine, argued that not reporting on the case "continues the erasure of Black women from the national conversation on race, police brutality and the right to safety."[4]

"Erasure" is a term, originally coined by scholars in academia, that has found its way into the mainstream. It signifies our society's indifference to specific people or groups—effectively rendering them invisible. Insofar as the society is concerned, their pain just doesn't matter. Erasure demands that we ask ourselves: "Whose stories do we tell?" "Whose pain matters?" "What, if anything, will move us to act?"

Erasure almost always affects the powerless: women, people of color, those without financial means. In the realm of science and research, for example, women's contributions have been erased so frequently and for so long that they even have a name for it: the "Matilda Effect" (named for a nineteenth century activist in the women's suffrage movement, Matilda Joslyn Gage). I have experienced the "Matilda Effect" in my own work, seeing my contributions minimized or erased while others—usually men or white women—take the credit.

Through it all, we have resisted. As with the Holtzclaw case, Black female journalists and activists have demanded that society pay attention to the ways in which crimes against Black women have been ignored and overlooked altogether. The #SayHerName movement elevates the victims of police brutality—Black women like Alexia Christian and Meagan Hockaday whose deaths have received only a small fraction of attention relative to the tragic murders of Eric Garner or Michael Brown. In the case of missing girls and women, California journalist Erika Marie Rivers launched *Our Black Girls* to highlight the fact that one hundred thousand Black women go missing every year. *One hundred thousand.* Yet our society engages with just a few of those stories if we're lucky.[5]

This isn't, to be clear, about competition. I am not comparing the relative horror of one tragedy with another. It is horrific and wrong when *anyone* dies at the hands of police. It's tragic when *anyone* goes missing. But this conversation is about justice. And fairness. It is about building a country that truly cares about all its people, no matter their race and no matter their gender. To do so, we must begin with acknowledgment: countless studies have shown that Black girls, women, and nonbinary people are hypervulnerable to abuse—especially in the context of our justice system.[6] Consider the facts: Twenty-two percent of Black women in the US have experienced rape. Forty percent of Black women will experience intimate partner violence in their lifetime. More than any other group of women, Black women are murdered at a higher rate. Among Black transgender and nonbinary individuals, 53 percent have experienced sexual violence and 56 percent have experienced domestic violence. In 2018 alone, at least sixteen Black transgender people were murdered.[7]

The bias throughout society, the bias that each of us carries, leads many of us to view Black women, as well as transgender and nonbinary individuals, as *deserving* of violence instead of victimhood. That blunts our response to what should be considered a national crisis. The casual discrimination that our society perpetuates against Black women, transgender, and nonbinary people contributes to a cycle: first, we are overlooked and ignored; and then, we are dehumanized in our victimhood. Our pain and suffering just don't add up enough to matter. When justice isn't served, the cycle repeats and continues uninterrupted. As a community, we are left exposed and vulnerable—preyed upon and unprotected by broader society.

Worse yet, the silencing of Black women, girls, and nonbinary people as they endure the structural violence against them leads to other devastating consequences, including the incarceration of the survivors. Often, when Black survivors of abuse defend themselves, they risk their own incarceration. Consider the case of Cyntoia Brown, who was only sixteen years old when she shot and killed a man who had picked her up for sex. She feared for her life when the forty-three-year-old man took her home and showed off his gun collection. The prosecution argued robbery and Brown, who was charged as an adult, was sentenced to life in prison. After an extensive clemency campaign, her life sentence was finally commuted after she had served fifteen years in prison.

As with all these tragedies, Brown's personhood was erased. The process that led to her incarceration portrayed her suffering as deserved.

The denial of humanity, endured by Black women, transgender, and nonbinary individuals, is a profound omission—it leads our society to neglect and hurt us.

We do our best to fight back against erasure as we live our lives and pursue our work and careers. As a journalist, I encounter erasure every day and routinely observe it in our politics.

For instance, here in my home state of Maryland, there is fawning coverage of white women in politics by local and national news publications. Yet routinely, the contributions of Black women are overlooked. State House Speaker Adrienne Jones, the first Black woman in that seat and one of a handful nationwide, sees her contributions routinely omitted from state or national press coverage. Alexandra Hughes, her chief of staff and another pathbreaking Black woman, has to fight to correct the record in the press—to ensure that the public is aware that one of their state's top leaders, who happens to be a Black woman, is delivering for her constituents.

Although infuriating, this phenomenon is not surprising. Culturally, Black women are expected to recede into the background and to serve others, without complaint, without acknowledgment. This is an important point. While many would like to see Black women marginalized in The Conversation or The Work, we're expected to support and serve, without fail. Black women fulfill a disproportionate number of service roles in the United States. We are everywhere.

UNDERESTIMATED, UNDERVALUED, AND ESSENTIAL

It should go without saying that Black women are central to the nation's economy. We are overrepresented in the nation's frontline jobs: we are the manicurists and pedicurists, maids and housekeepers, graders and sorters in agricultural work, home health aides, nursing and medical assistants, cashiers, teachers, teachers' assistants and childcare providers, waitresses, retail workers, and on, and on, and on. Black women dominate the ranks of the often-nameless faces that check out your groceries, prepare your order, attend to your health when you're hospitalized. Think about this: if a job has nonstandard hours, fewer protections, and less flexibility, then Black women are most likely to occupy that job. As the saying goes, no good deed goes unpunished. Among all full-time workforce participants in 2018, Black women earned just 63 cents for every dollar earned by white men (as compared to the 78 cents earned by white women).[8]

Stepping back from specific roles and responsibilities, we participate in the US workforce in high numbers. We work, *a lot*. As parents, Black women are, more often than not, working parents *and* the breadwinners for our families—a notable and important gender reversal. At work, we keep the country running and at home, we likely shoulder the responsibility of keeping our households afloat. Half of Black female workers are mothers, and more than two-thirds of working Black mothers are single. We are the future: in 1974, a married couple headed 84 percent of *all* families with children.[9] By 2017, that number fell to 66 percent as the number of single-parent and "nontraditional" households increased.

Black women also motivate a huge share of America's small business and entrepreneurial activity. Among Black-owned businesses, Black women make up 46 percent of Black entrepreneurs, one of the highest percentages of female small-business owners of any segment.[10] We represent one of the fastest-growing segments of American entrepreneurs. Throughout the economic collapse created by COVID-19, small businesses led by Black women, relative to every other demographic group, dramatically expanded their services and offerings to steer their communities through the crisis.[11] Unsurprisingly, when it mattered most, the Black female entrepreneur doubled down on ourselves and our communities. Nevertheless, the country refuses to double down on us. Just 0.64 percent of venture capital flows to Black women.

Here's an excerpt from "The Hidden Help: Black Domestic Workers in the Civil Rights Movement":

> During the 1960s, nearly ninety percent of Black women in the South worked as domestic servants. While much has been written depicting the dehumanizing and exploitative conditions in which they lived, their contributions to human rights . . . has either been undocumented or documented quite minimally . . . Domestic workers were an unpaid community of organizers who helped mobilize, coordinate and inspire boycotters and demonstrators. The willingness of domestic workers to house visiting student freedom riders, to make protest signs, and present as the largest group of bus boycotters represented the recognition of formal constitutional rights.[12]

For Black domestic workers, their behind-the-scenes leadership was the only role available to them. Much like Fannie Lou Hamer, who lost her job when she registered to vote, Black domestic workers could least afford to pay the price to exercise their right and yet they did so anyway. Along with students, but far less celebrated or known, they constituted the bulk of the grassroots leadership in the civil rights movement. They canvassed their local communities, registered voters, and nurtured all who participated in the movement—doing thankless domestic work at home and operating behind-the-scenes on the front lines of activism.

Scores of Black women, like Aurelia Young, housed and fed students of the Freedom Rider movement after they were released from jail. They fed civil rights workers and attorneys in their homes—critical labor that has gone unrecognized in our celebration of this critical chapter in American history.

As Mayor Lightfoot said, for Black women in particular, "*The cavalry's not coming to save you.*"

She's right. And there's one big exception. Black women *are* the cavalry, we always have been. After all, sisterhood is our superpower. Lightfoot talks about this power:

> We are very much in contact with each other and supported this week watching what's happening in New Orleans with the hurricane. We have all reached out to LaToya to give her love, to let her know that we're here for us, and to offer any kind of support that we can. Same thing with Muriel Bowser, she has gone through hell and back during the Trump administration, we were all there cheering her on and picking her up when she needed a lift. So, we are there for each other because there are very few people who

walk in our shoes and really walk in our shoes, not just any mayor, but a Black woman mayor in big cities across this country. This is a very challenging time for all of us to be in leadership. And again, I just can't tell you, I am actually the youngest in tenure of the mayors. I think I am the oldest as you can tell from my gray hair. But I am the youngest in tenure and these women have just been a lifeline for me. I have great, great admiration for every single one of them. And we steal from each other. We call each other for advice. We're united in making sure that we're doing the best that we can for the residents of our city because we know that for many of our folks, generations of disinvested generations are not having a seat at tables of power. We're upending the status quo.

Mayor Lightfoot's parents, who grew up in the segregated South, raised her to be tough and resilient. And she is. The mayor took a very tough stand against the "overwhelming whiteness and maleness" of the media when she decided to grant an interview to only journalists of color. This move was hugely controversial but lifts up a powerful example of sisterhood and solidarity. As a Black journalist who has struggled for her seat at the table, I appreciate the rare opportunity to have my voice and perspective elevated in reporting. After three decades of reporting, I can safely say that it almost never happens.

CHAPTER 3

Walking the Tightrope

You may be the first to do many things, but make sure you're not the last.

Vice President Kamala Harris

**LESSONS FROM OBAMA,
NEW CHALLENGES FOR HARRIS**

In the summer of 2009, President Obama saw the biggest single decline in his approval rating of his entire presidency. In the first half of the first year of his presidency, Obama enjoyed the support of a large majority of the American people as he steered the country through a once-in-a-century financial and economic crisis. Despite the troubled state of the country when he first assumed office, Obama still basked in the warm glow of his election as America's first Black president. The press and public exalted in the postracial America that Obama's election appeared to signify. After all, the Harvard-trained lawyer who now occupied the White House was so different from his predecessors. Maybe he was the one to guide the country to a better place—a destination far past the racial division that defined America throughout history.

But that goodwill evaporated in a moment during one fateful press conference. In July 2009, a white police officer arrested storied Harvard professor Henry Louis "Skip" Gates outside of his home in Cambridge, Massachusetts, when the officer mistakenly assumed that Professor Gates, a Black man in his late fifties, was breaking into his own home. The incident grabbed local headlines but went national when President Obama was asked about it and he responded that the Cambridge police sergeant, James Crowley, acted "stupidly" when he pursued the arrest.

Obama's comments proved fateful because he immediately confronted national outrage. The president wrongly assumed that the casual racism portrayed by the officer in the stop was obvious to any fair-minded observer. After all, who arrests an older Black man at his own home for trespassing? Nevertheless, many white Americans felt that the president was not sufficiently supportive of the nation's police force by failing to sympathize with the officer's position. For them, when Obama centered Professor Gates's race as the reason for the senseless arrest, President Obama was simply playing the "race card." Obama learned then that there would be little truth telling when it came to race during his presidency.

He paid a price for the lesson: in one fell swoop, Obama saw his approval drop seven points—the largest single drop of his presidency—and he never recovered with certain white voters.

Throughout his presidency, Obama was captured by paradox. He was obviously Black and a historic figure precisely because he was the first Black president. Yet many white Americans expected Obama to be, in every respect, postracial. From his remarks and comments to his actions as a policy maker, they didn't expect the charming, well-spoken political prodigy from Illinois to actually name and acknowledge race. For these Americans, Obama's mere presence in the White House was proof that racism was dead in America and, therefore, his presence in the White House should only affirm that belief. For Obama to contradict that belief by calling out racism as a current and persistent problem proved a bridge too far for many white Americans. As it dawned on white America that Obama did not function to salve "white guilt," many turned on him.

What a turn it was. A seven-point drop in approval is not sustainable in politics. Too many of those kinds of drops and you'll find yourself out of office, pronto. The rapid change in the president's behavior showed that he had received and had heard the message, loud and clear. The president quickly and thoroughly walked back his comments and the now-infamous "Beer Summit" leaned on then Vice President Joe Biden to help clean up the public relations mess. Ever cautious and acutely aware of his tenuous position, the president would steer clear of any discussion of race until the killing of Trayvon Martin three years later, when he remarked that the young man "could have been his son." But at no point in his presidency did a racial equity agenda or anything close to it dominate the president's policy and political strategy—his calculation was that he couldn't afford to operate that way.

This episode early in his presidency taught Obama about the dilemma that Black people face, daily, at the highest levels of their profession. Think about it: despite the progress of Black people across society, even the most powerful man in the world found himself captive to America's difficult and often toxic racial dynamics.

And as difficult as it was for President Obama, the most famous and powerful Black man in the world, it is *more* challenging for Black women at the top of their game.

I would know. For years, I was the sole Black woman focusing on urban America in the White House pressroom. While I always spoke my truth, I couldn't always speak my mind. I learned early that if you cross a line by making someone, particularly a white person, uncomfortable with your questioning, you could see your career held back for years. If you are viewed as "angry" or "hostile" in your demeanor, then you could see your career derailed altogether. As Obama discovered, if you are not attuned to those cultural threats and, in turn, you do not remain viable, you cannot lead because they'll push you out of the room. If you cannot lead, you cannot change things. And ultimately, you're in the room in the first place to try to make a difference.

After all, it's a tightrope.

In journalism, especially when the beat is politics, a journalist lives and dies by their ability to cultivate sources and produce stories. The essential contacts for a journalist—the press secretary, key administration officials, party insiders—reflect the society. Those essential contacts share many of the same blind spots and biases that have affected Black women in society for centuries.

What do you do? You learn to walk the tightrope.

When society, particularly men, objectify our beauty as lascivious, lewd, and hypersexual—the ancient jezebel trope that is as old as it is *old*—we stand in our beauty and grace to channel our strength. We hold our head high and find solace in other Black women who see us for who we really are. We resist the stereotypes and project our truth with dignity and in solidarity.

When society tries to silence us by shutting down our voice, we speak our truth and then organize around it—we mobilize our allies. Together, we amplify. We refuse to cower. We find our strength in both our numbers and our individual experiences.

When society tries to minimize us—whether it is at work or on the front lines of activism—we take up space anyway. We insist on our presence being felt. Even if we are quiet, we are here. Through our presence, we push back.

No matter what room we are in, we fight for those outside of the room. We know that we are here for a reason and that reason lives in every person, past or present, who has found themselves marginalized and excluded. We're here to fight for what's right.

All the while, we resist, we get loud, we take up space, and we fight without going *too* far. Our response is almost always a controlled rage. *We've learned to play the game while changing it. That's walking the tightrope.*

Those of us who are among the first in a particular space or even an entire profession pay a particular price. To paraphrase our vice president, when you're the first, your responsibility is to make sure that you are not the last. As President Obama discovered and as Vice President Harris is discovering, you can't take as many risks or move as boldly as you would like lest you risk slamming the door behind you.

The daughter of immigrants, one Indian and one Jamaican, Vice President Harris has endured insinuations about her talent and career throughout her entire time in public life. Like Obama, Harris navigates tricky racial politics. On every issue, she must weigh her words carefully for fear of them being misconstrued or racialized in a polarized society that struggles to confront and unpack America's ugly history of race. But she must also account for her gender: her intelligence, poise, and, indeed, her beauty have made her a target from the beginning. Her critics, her rivals, and even her friends have either deliberately or inadvertently delegitimized her accomplishments on account of whom she has dated or where she has socialized. Her early career was marked with nasty murmurings that implied her early, precocious success in politics was the singular result of whom she had dated.

Despite the tropes and stereotypes, Harris has walked the tightrope as successfully as any politician in American history and is now the highest-ranking female federal official in the history of the country. She has resisted the incessant commentary about her looks by carving out her style, confidently wearing either Chuck Taylors or heels.

As the world found out during the Brett Kavanaugh US Supreme Court confirmation hearings, when Christine Blasey Ford endured cynical attacks on her character and motivations, then Senator Harris wasn't afraid to get loud and ask the nominee difficult questions. She organized both her Senate peers and her grassroots allies—her voice was incisive and in the lead, but she was never alone.

Throughout her presidential campaign, she bounced back from failure by reading the political moment and building an alliance with Joe Biden—taking up space in a consequential race in a new and powerful way. Instead of going away to lick her wounds, she transformed and emerged more powerful and more focused.

Now as vice president, she is acquiring a reputation as the administration official most interested in building authentic and deep ties with the progressive base. She's fighting for those not in the room, even—or especially—when it's not easy.

And as President Obama discovered, Vice President Harris is learning all the ways that it is not easy to be the first. As of this writing, it is not even a year into the Biden administration and Vice President Harris holds almost all of the president's most challenging portfolios: immigration, police reform, voting rights, COVID-19, workers' rights, and bridging the digital divide.

These controversial issues have not won her any friends. And yet, it's the price of being first, it's the job.

. . .

President Biden is lucky to have Kamala Harris by his side. Because she is The First, Vice President Harris has unique strengths and particular burdens. The president understands this and leverages her unique profile to benefit the overall success of the administration.

The truth is that President Biden has relied heavily on Black women more broadly since the start of his campaign. Without Black women in his inner circle, without the enthusiasm of Black women at the voting booth, and without the heroics of legendary organizers like Stacey Abrams, including her legion of followers, the dedicated activists—some known, most unknown—who did the hard work of registering, turning out, and protecting voters, Biden would not be president. Period.

Atlanta Mayor Keisha Lance Bottoms has played an especially vital role in supporting President Biden. She helped presidential candidate

Biden survive a viciously competitive primary that was almost over for the future president before it started.

Bottoms was first on the scene to support Biden when questions about his commitment to civil rights dominated the conversation. The skepticism about his commitment to civil rights increased during that first presidential candidates' debate in June 2019, when then Senator Harris questioned Biden's efforts to block the busing of Black students into predominantly white neighborhoods in the 1970s when he was a US senator.

Harris's remarks cut Biden, deeply. For months, he hurt to his core.

"You know my heart," Biden said to me in July 2019 on a stage at the NAACP Presidential Candidates Forum, where I served as moderator. He repeated himself; this time his tone was vulnerable—even wounded.

"You know my heart."

His expression was pained. This man hurt over the insinuation that he had opposed racial equality at an important peak of the civil rights movement. And he didn't let it go. Backstage in the greenroom, Biden shared with me that he and Harris were friends. He had long been supportive of her political career, and then California Attorney General Harris was close friends with his late son, Beau Biden.

As I look back, I now understand the situation in which Senator Harris rode into the Detroit convention center in her Black SUV, staying inside until it was exactly time to get onstage and take questions from me. Politicians, staffers, and members of the press milled around to meet and greet before the debate, but Harris stayed away in an effort to avoid any potential awkwardness or uneasiness that might follow during the debate. I mingled with an impressive slate of presidential candidates: Elizabeth Warren, Pete Buttigieg, William Weld, Amy Klobuchar, Beto O'Rourke, Joe Biden, and Cory Booker. But it was Kamala Harris, the Black woman who would go on to break the color and gender barriers to the vice presidency, whom I didn't see. The disappointment and surprise felt by many of us in the press when she didn't get out of the SUV was too powerful to ignore.

As a journalist and a Black woman, I believe that Senator Harris was right to bring up Vice President Biden's record on civil rights. We've seen what happens when a politician is not held accountable for past thoughts, comments, and actions on matters concerning people, particularly the most vulnerable among us. While Harris was correct in bringing up the busing issue, I am glad that the fracture in that relationship was healed. No matter how it happened or what it took to heal it, that relationship has ascended to the highest heights. Trust is *the* major factor in Joe Biden and Kamala Harris serving as partners as they lead this nation.

Biden's actions should also be put in context by his recent record. After all, he served our nation's first African American president with dignity and unfailing loyalty. Biden jumped into the presidential race because of the deadly race riot in Charlottesville. It is hard not to imagine that Harris's pressure pushed Biden further in his commitment to racial equity. Undoubtedly, the allyship of Mayor Bottoms helped save Biden's candidacy. Bottoms, a prominent Black woman, stood up for Biden as the busing conversation swirled and did not let up in the press for weeks. It dominated the headlines and it affected the polls: Black Americans were not feeling Biden or his attempts to become the forty-sixth president. Bottoms helped blunt the impact of Harris's remarks at the debate and the subsequent fallout.

Biden emerged from this controversy a different man and a better leader. As of this writing, President Biden is leading the most ambitious effort to advance racial equity in the nation's history. To paraphrase the late poet and activist Maya Angelou, when you know better, you do better.

To reward her allyship, Bottoms's name was tossed about in the vice-presidential ring as a potential running mate for candidate Biden. She belonged to a cohort of Black women to vie for the position, a group that included Susan Rice, Karen Bass, Stacey Abrams, and Harris. Looking back, I was skeptical that a Black woman would be chosen to be the Democratic vice presidential candidate. After all, as Black women, we are used to being used—discounted, dismissed.

Biden had said he would nominate a Black woman to the US Supreme Court (a promise he kept in early 2022, when he selected Judge Ketanji Brown Jackson, who was confirmed as a Supreme Court Justice in a bipartisan vote). However, he never promised the vice presidency to a Black woman. Talking to a source very close to Biden, I did not expect a Black woman to be on his presidential ticket because "he does not work that way." Keep in mind, there were impressive white women on Biden's mind too. Some of them were among the most recognized and accomplished figures in the presidential primary season: senators Elizabeth Warren and Amy Klobuchar.

Behind the scenes, I had honest talks with political commentator Bakari Sellers, who said it was a done deal: it was Harris. Ultimately, Harris and Rice were the top picks for the job because of their experience. However, I couldn't get my conversation with Biden in Detroit out of my mind. As for Rice, the embassy massacre in Benghazi, Libya, in which the US ambassador and three others were killed, was disqualifying.

I will never forget that day at Andrews Air Force Base, where I observed the peaceful transfer of the remains of those who had died in that deadly attack. I stood there as a pool reporter and watched one of the most solemn moments in our nation's history. Former Secretary of State General Colin Powell was also there. On the Sunday news shows, administration officials all said that the tragedy in Benghazi was the result of an intelligence failure. But the news shows blamed Rice for the crisis. Once again, it was a Black woman who bore the brunt of the fallout. That difficult experience may prevent her from being elevated to a Senate-confirmed position or an elected office.

Sellers was right. If I had bet money, I would have lost. The announcement came and it was a Black woman, Kamala Harris. She was more than qualified and the moment of racial reckoning called for transformation from a transformational figure. That was Kamala Harris, now Vice President Harris. As a Black woman who wants examples of greatness for our younger generations, I find it amazing to see her. It is important for the Black women of today to know that we are now seen.

With the good comes the not-so-good.

After the days and weeks of wrangling, Biden was confirmed as the president-elect, and Mayor Bottoms was ostensibly positioned for greater things. However, the Biden transition team's decision about her appointment did not sit well with her. It was time for Bottoms to walk the tightrope.

She was not given a cabinet seat. Rumor had it that she was up for heading the US Department of Transportation. As head of government in Atlanta, Bottoms oversaw the largest airport in the nation. She had worked with top-tier Fortune 500 companies that called Atlanta home and headquarters. Still, presidential transition officials juggled names and seats to find the right fit. Ultimately, that seat went to former presidential candidate and former South Bend, Indiana, Mayor Pete Buttigieg. Next, Bottoms was in line for the role of Small Business Administration administrator. However, Bottoms decided she was more than qualified to have "Secretary" in front of her name. She decided that she was out of the running. Many close to Bottoms resented the administration's attempts to sit her at the "kids' table."

It was a bold move to turn down the role of SBA administrator. The conventional wisdom in the Beltway is if you are asked to serve, then you serve. While it is not unheard of to turn down a position in government, it's rare. Bottoms's thought process went against the convention. She said "no" and never looked back. The move to Washington, DC, and the cost of uprooting her family just weren't worth it. She decided to stay in Atlanta and contemplate a potential reelection run. That decision was short-lived. It was upended by the increase in crime nationally and in her hometown. She did not run for reelection.

Bottoms is considered a sweetheart in politics. She works with a firm hand and kind smile. She enjoys support from so many sectors, including that of her sorority Delta Sigma Theta Sorority Inc., one of the most powerful sororities in the nation.

. . .

Black women have always grappled with the expectation that they should operate in the background, that somehow we are expected only to serve and be subservient. While we have made huge strides in political representation, especially in the last five years, that expectation continues to weigh down Black women, even as we climb to reach the highest echelons of political power.

It's critically important to evaluate in context the contributions of the most powerful Black women in the United States. In just a few short years, Black women have gone from being the unsung heroes of the Democratic Party and the civil rights movement to playing a starring role in both—from the vice presidency to the Black Lives Matter movement, Black women are *powerful*.

It's all so new, so new in fact that we still navigate double standards. Society is still learning to accept our power and to create space for us in which we fully wield it. We are expected to be strong *but not too strong*. We are expected to play a prominent role in national life *but not too prominent*. We are expected to hold the powerful accountable and to speak truth but not speak too forcefully or be too critical, lest we earn unwelcome criticism of our own. Harris is navigating this difficulty correctly as the administration slow-walks its commitment to voting rights—perhaps the defining issue of our time. Vice President Harris is on the record[1] for pointing out the vital importance of voting rights and yet, there is only so much she can do before she risks slamming the door behind her.

Here's an excerpt from "Recognizing Workplace Challenges Faced by Black Women Leaders":

> White women are stereotypically seen as communal: pleasant, caring, deferential, and concerned about others. Their leadership challenge, therefore, is to avoid being seen as so communal as to be an ineffective leader without being seen as so agentic as to be unlikable. Black women face a very different challenge. They are not stereotypically seen as communal but rather as assertive, angry, and "having

an attitude." Their challenge, therefore, is to avoid being seen as so angry or assertive as to be unlikable without being seen as so subservient and compliant as to be lacking in strength and independence.[2]

This is an important point. *Both* white women and Black women face "lose/lose" scenarios in the workplace. Nevertheless, a Black women's predicament is more acute. If a white woman is viewed as too communal to lead, she may be ineffective but she's still likable. In this analysis, at least, the white woman is unthreatening if impotent. But if a Black woman is viewed by her colleagues as angry, she is not only viewed as untenable as a leader but also as unlikable and, quite possibly, hostile or even a threat.[3]

Race plays a particular role too. In the workplace, white women and Black women both struggle under the weight of gendered norms. For all our progress, society still affirms and skews toward masculine behavioral norms. This effectively levies a behavioral tax on all women as they navigate a society that doesn't always respect, elevate, or reward feminine models of contributions in the workplace. But Black women must also conform to dominant white behavioral norms too. As a result, Black women are under pressure to dress a certain way, wear their hair a certain way, or speak a certain way—all in an effort to appear less "Black" or "ethnic." Survey after survey shows that even the most successful Black women, such as graduates of Harvard Business School who go on to huge careers in corporate America, report that they struggle "to be themselves" at work. Once again, Black women walk a tightrope that balances fitting in at work with being authentically themselves.[4]

The bias that Black women face in their work adds up to hold them back in their careers. Below are findings from an article[5] that describes how Black women leaders are evaluated more negatively than Black men, white women, or white men in a struggling organization because of the "double jeopardy" effect:

- *On a seven-point scale, men are perceived as more effective leaders than women,* with scores of 4.52 to 4.11 points on average, respectively; and white individuals are perceived as more effective leaders than Black individuals, with scores of 4.44 and 4.17 points on average, respectively.

- *On a seven-point scale, women were also perceived to be less typical leaders than men.* Black individuals are perceived as less typical leaders than white individuals.

- *Under conditions of organizational failure, Black women are evaluated more negatively and as less typical leaders compared to both Black men and white women.*

Black women in leadership are easy scapegoats when things do not go well at work because they are almost always viewed by their peers as "atypical" leaders. Because of bias, Black women in leadership do not fit the "norm." Therefore, when things go poorly at work, many default to the mental shortcut to blame a Black woman because they are the "atypical" colleague at work. That white men always score more favorably, even when the organization is struggling, affirms the power of fitting the "norm" of "typical" leadership: white and male. That white women and Black men score in between Black women and white men underscores the privilege of gender and race that white women and Black men enjoy, respectively.[6]

That doesn't mean Black women aren't asked to lead. Indeed, Black women usually find themselves in less desirable leadership positions—taking on the hardest assignments, putting themselves and their careers in jeopardy while doing so. Here's an excerpt from "Leading While Black, the Experience of Black Female Principals":

> Black female principals also tend to be relegated to the most segregated and poorly resourced schools in the country. These discriminatory practices create unique challenges for these women and their ability to thrive in their jobs as school leaders.
>
> According to the National Center for Education Statistics, just 10 percent of principals are Black and 80 percent of public-school teachers are white. These demographics highlight the isolation Black women principals often experience, as well as the structural nature of their exclusion.[7]

When Black women lead and things are going well, we usually struggle to be recognized for our work and our achievements. The contribution of women in the workplace, especially Black women, is largely invisible and taken for granted. This is a phenomenon that unfolds, unchecked, in the world of work because the contribution of women is usually conflated with assumptions about "what we're naturally good at," "interested in," or "passionate about." Here is an excerpt from "Women Leaders Took on Even More Invisible Work During the Pandemic":

> Women are not rewarded for capacities and concerns deemed to be intrinsic. Therefore, when a woman manager provides team members with emotional support during a time of societal crises,

it can be overlooked as "caretaking" instead of being recognized as strong crisis management. When a Black woman manager hosts a panel on anti-racism in the wake of racial violence, she can be applauded for her "passion" but not rewarded for her time, leadership, or DEI acumen. . . . Since recognition and reward are the markers of valuable work, that women leaders' efforts are going unnoticed and unrewarded effectively renders it low status.[8]

On the biggest stage in the world, Kamala Harris has experienced the hard truth of working and leading while Black and female. Her experience mirrors the journey of Black women everywhere toiling to climb to the highest altitudes of our respective professions: it is not just lonely at the top; it can be hard to breathe at all.

Section II

The
Price

WHAT WE ENDURE,
HOW WE OVERCOME

CHAPTER 4

Our Fight

Would you want to trade places with me?

April Ryan

THE WARNING

I don't know how it is that the nation has forgotten what took place in Charlottesville. I was invited to the Twenty-Third Annual Virginia Festival of the Book to discuss my book *At Mama's Knee.* The timing could not have been better. Only months into the Trump presidency, I was already wrestling with a growing uneasiness: this isn't normal. The new administration found it too easy to lie, to mislead, to obscure the truth. Unlike his predecessors, our new president did not govern for all America. His hostile tone and a fast-and-loose relationship with facts demonstrated the truth that he arrived at 1600 Pennsylvania Avenue to fight for only some Americans: the "forgotten" white male.

I experienced the hostility of the new administration every day at work. Under this president, in just a few short months, the White House pressroom changed. Gone was the respectful tone and professionalism that President Obama and every prior president in the modern era established and maintained. This president was adversarial. In the very first press briefing, we debated the actual size of the crowd at the new president's inauguration. From that moment on, it became increasingly apparent that this administration viewed you as either friend or foe.

Just two short months into the new president's term, I found myself exhausted. The 2016 election had been grueling. It was a divisive and ugly campaign, and covering candidate Trump demanded an unusual level of vigilance from the press corps that was extraordinary and taxing, even by presidential standards. Daily, Trump broke norms and news. His appetite for coverage, any coverage, was insatiable. His recklessness and narcissism spilled over into the chaotic first days of his presidency.

As one of the few Black women in the press corps, I was particularly affected by the behavior and culture of this very different White House. From jump street, my identity made me a target. My reporting has always focused on the human rights and daily struggles of everyday Americans left behind in a country marked by growing inequality, and the nature of my reporting only intensified the hostility from the administration. Only a few months into a new presidential administration, I was left feeling empty.

Do I really have four more years of this?

My exhaustion was real and I felt guilty because of it. As a member of the White House press corps, I was blessed to participate in the pinnacle of my profession. Even so, my gender and race invited a near-constant barrage of hostility because for all her progress, America still grappled with the sickness of discrimination and bias. Whether you are a reporter stationed at the White House, a nurse serving in a hospital, or a teacher instructing in a classroom, in America you are a still a Black woman—uniquely burdened and specifically targeted by a society and a system that struggles to fully embrace your humanity.

I also experienced guilt because I had sacrificed so much to reach this point in my career. The life of a reporter, especially at this level, is not an easy one. On the campaign trail especially, you spend countless days and nights away from your home—distant from loved ones and disconnected from your community. Black women in journalism have to give up so much to establish themselves in the profession. Journalism isn't unique. It's well-chronicled the extent to which Black women pay a higher "price" for their achievements—from our education to our career to our health and even our families, we "trade" more to get a little less than everyone else. The life of the Black woman is partly characterized by the endurance of inferior treatment, distorted expectations, and an uneasy sense that we'll never fully belong.

For much of my life, to achieve my goals, I have compartmentalized the pain of mistreatment. To refashion the old saying about working "twice as hard to get half as far," I have worked 400 percent harder to reach the pinnacle of my profession. Instead of savoring this stage of my career, I felt only heaviness and the burdens of responsibility in the Trump White House: If I didn't ask these questions, who would? If I didn't represent our people in that pressroom, who would? Who could?

Promoting a new book offered a reprieve from the grind. Even in normal times, I looked forward to interviews and speaking gigs. Publishing is a blessing in my life. It provides me the chance to break out of the DC bubble and escape the twenty-four-hour news cycle that dominates our political culture. Books empower their authors with the opportunity to connect with their readers on issues important to both of them. Directly connecting with your audience in person is a little like drinking a glass of cold water on a hot summer day. It's refreshing. It's nourishing. And unlike political journalism, which obsesses, perhaps to a fault, on "who is up and who is down," a good book invites conversation and facilitates deeper understanding for all involved. After what felt like a lifetime in the bubble of the White House press corps, I was ready to step back from the pressures of my day job and reconnect with America and her people.

In late March 2017, I accepted a speaking opportunity at the Virginia Festival of the Book in Charlottesville. For a couple of days, I would leave my worries behind and my two teenage girls at home and supervised. I took off.

Charlottesville is a hike from my native Baltimore. But once I get beyond the traffic migraine that is the Beltway, the drive is absolutely beautiful. As I drove deeper into Virginia, I bathed in the natural beauty of the Shenandoah Valley. My breathing and heart rate slowed. I could even feel my cortisol levels starting to fall. I caught glimpses of chestnut and red oak trees peaking above the horizon as I meandered up and down long and winding roads into the small college town.

Charlottesville is beautiful in late March. Spring is in bloom and the town is alive with activity. As I strolled down Charlottesville's cobbled sidewalks, I soaked in the moment: students bustling up and down the sidewalk, small quaint shops dotting Main Street, and majestic ancient trees, just beginning to blossom, in the median to divide the roads.

I am not the biggest shopper. After all, who has the time? But on this day, with a few hours to kill before my talk, I stepped into a quaint jewelry shop that I quickly learned had been there for years and years and years. I struck up a friendly conversation with the owner, a kind and chatty white woman, who steered me toward a beautiful silver-and-leather bracelet.

The owner was proud to call Charlottesville home. She was a lifelong local who gushed with excitement about the book festival because it marked the beginning of the tourist season as out-of-towners poured into the cozy downtown for their summer shopping. I perused as we chatted before finally buying the first piece of jewelry I had seen. After saying good-byes, I left for a nearby building to give my talk; I was adorned with my new jewelry and full of warmth from having made a new friend.

Before a large and diverse audience, I spoke about *At Mama's Knee*, my memoir about raising two Black daughters as a single mother. In the book, I discuss navigating the racial tragedies and traumas that echo those of generations past: Trayvon Martin, Michael Brown, Emmett Till. So much about what we understand about race is illuminated by the families who raise us. Without dialogue and exposure, many white families never fully understand the struggle of Black Americans. As much as our country has changed for the better over my lifetime, the fundamental struggles of being Black in America remain largely unchanged: the income gap separating white and Black families is as large today as it was in 1968 (around the time when America declared a "war on poverty"—a war, I would wager, that we have lost). During the same time span, the wealth gap has grown even larger. Black Americans are sixteen times more likely to grow up in families with three generations of poverty than white adults. Sixteen times![1]

For all our progress, America requires—hell, demands—structural change, which will follow only after a shift in our politics. That change will never happen without achieving deeper understanding among and between white and Black people throughout America. Without healing and reconciliation, how will we move forward as a country?

After I wrapped up my talk, I greeted a line of audience members, Black and white, young and old, who wanted to ask a question or express their opinion about something I had written or said. This is my favorite part. I love these moments. Books create opportunities for rich dialogue and small powerful moments that never leave you.

"I give my daughters the exact same advice," one Black mother said, clutching my hand tightly in hers. "I laughed out loud when I read about your daughters," one young Black father shared with me. "Your daughters remind me of my own. The next generation always takes their privileges for granted. Indeed, that's their privilege. It's a reminder of why we fight. Even if it's a little frustrating, at times."

"I learned so much reading this, thank you, and thank you for coming to Charlottesville," said a white woman. Her eyes were kind and her tone sincere as she tightly held my hand.

As the room slowly emptied out and the technical staff started to prepare the room for the next session, a large white man milled about in the back of the room. He wore a faded green T-shirt and blue jeans. The number of people in the room grew smaller and smaller, and I took more and more notice of him. The same uneasiness I felt in the Trump White House returned. I felt a small knot in my stomach that hardened into an ache when I remembered that I was alone on this trip. I had forgone the security protocol that I had been recently forced to use. After all, the death threats and threats of violence against me and my family exploded after the country elected President Trump.

Political reporting, for Black journalists in particular, started to grow more dangerous after we put a Black man in the White House.

After 2008, our politics radicalized seemingly overnight. That was not Obama's fault; rather, a growing extreme element of Americans took offense at a Black family living in the White House and their agitated, aggressive, hate-filled emails arrived in my in-box daily. That trend only intensified after Donald Trump entered the White House. He encouraged and enabled the derangement through crafty dog whistles—nods and winks.

"Ms. Ryan?" the large white man finally said to me as the last guest left the room. His voice was urgent. He came close and, in an instant, we were alone, standing closely together in a cramped doorway. The event organizers had cleared out, at least temporarily.

"Yes," I replied, tentatively. I was five foot three, and his six-foot-plus frame loomed large over me. "I don't agree with what you said," he began, clearly agitated. His eyes glistened with emotion, and his voice was barely restrained. He fidgeted with his hands and shuffled his feet. "I was diagnosed with a learning disability," he continued, his voice growing louder. "And I overcame it. I beat it. And I don't like the idea of giving away anything to people when they can overcome just like me." He was yelling now. His implication was clear: He didn't care how bad Black people had it in America. If he could rise up, then Black Americans should be able to overcome their circumstances too.

"Your talk, it's just not fair," he continued. The man was offended by my suggestion that our government should help the vulnerable and the historically marginalized as a way of leveling the playing field. His face drew close to mine when I finally found my voice.

"Would you trade places with me?" I asked, looking him directly in the eyes.

"What?" His tone was confused and a bit exasperated.

"Would you trade places with me?" I asked again, this time louder and more insistent.

He huffed and then snorted, "You know the answer to that." He was angry now. I tensed up. The man stormed past me, heading for the exit and leaving me alone in the room that just moments earlier was filled with warmth. Now cold, my heart beat through my chest, my breathing was ragged, and a lump had formed in my throat. This is a warning, I thought, shaking my head in disbelief. A foreboding feeling swept over me.

WHEN THE WORLD DOESN'T LOVE YOU BACK

Black women are not given grace.
Christian Nunes, president of the National Organization for Women (NOW)

Rarely do I hear that truth spoken. Ever since, her words have haunted me.

What is grace? Grace can be a verb, an action of goodwill or mercy. On a deeper level, divine grace is the forgiveness we receive because of our humanity. God forgave Jesus his sins because he was human. To deny Black women grace is, quite simply, to deny our humanity.

Yet America denies Black women grace, every day. Every Black girl learns that the very first battlefield is the one in our mind, the lived experience in a society that struggles to value our humanity. From an early age, we learn to navigate. We learn to adjust. We learn to live with and, at times, confront the tiny voice in our head that whispers to us that we are not worthy—that we are not enough. That tiny voice is amplified all around us and reflected back to us through the culture and by our peers. At work, at home, in the world, we learn to resist the notion that, somehow, we are lacking.

We have fought for our humanity for centuries. Before Black women came to save the world, we had to learn to save ourselves because the evidence is clear: "Black women are not given grace."

In 2013, the Center for American Progress published an extensive report on the "State of African American Women in the United States."[2] Considering Black women as a population, the think tank evaluated key indicators in health, education, entrepreneurship, economic security, and political leadership.

It's a complicated story for Black women, who represent 13 percent of the female population in the United States.

On the one hand, Black women confront considerable challenges: Of all groups of women, Black women suffer from hypertension at the highest rate (46 percent compared to 31 percent and 29 percent for white and Brown women, respectively). Black women suffer from a higher mortality rate from breast cancer than white women, despite the fact that incidence of breast cancer is higher among white women. The mortality rate of Black women suffering from cervical cancer is twice the rate for white women. African American women represent 65 percent of new AIDS diagnoses among women.[3]

Black women are four times more likely to die from pregnancy-related causes, such as embolism and hypertension, than any other racial group. African American women have the highest rates of premature births and are more likely to have infants with low or very low birth weights. African American infants are more than 2.4 times more likely than white infants to die in their first year of life.[4]

These statistics represent a failure of systems that are not designed for Black women.

From the research and study of health issues to the training and development of practitioners to the people who deliver health care, Black women see themselves underrepresented—and at our significant peril.

Health conditions that disproportionately impact Black women—uterine fibroids, for instance—receive only nominal investment from the government. On the other hand, conditions that disproportionately affect white women see several times the government investment than those conditions that disproportionally affect Black women, even if the number of white women affected is far lower than the number of Black women affected. Further upstream in the process that determines which medical problems are ultimately addressed, Black women are severely underrepresented in clinical trials that require consent and are overrepresented in studies that do not require consent. Moreover, Black women are significantly underrepresented in key biomedical research data sets (in which people of European descent are overrepresented).[5]

The particular challenges of Black women in the health care system are rarely addressed in medical schools and training institutions around the country. Indeed, only very recently has there been an effort to reflect Black women in the content and illustrations in medical textbooks. The state of the health care workforce is defined by inequity too. Fewer than 3 percent of all doctors are Black women. Just 6 percent of all nurses are Black women.[6]

It is not surprising that Black women are suffering and dying at the highest rates among women.

We face similar headwinds in our pursuit of opportunity, from our experiences with education to the world of work. Of all groups of women, Black women are most likely to enroll in higher education and we complete our education at far higher rates than Black men. However, we still confront barriers to completing our degrees and even with our momentum in education, our value in the labor market is still diminished. Black women make only 64 cents on the dollar compared to white men (white women make 78 cents on the dollar). Black women make 10 percent less than Black men.

Despite these challenges, we refuse to give up. Of all women, Black women participate in the labor force at the highest rates. Even when COVID-19 drove women out of the labor market in droves, a phenomenon that was especially punishing for Black women, we still participated in the labor market at the highest rates—a trend that holds for Black moms as well.[7] Moreover, we are increasingly powerful and entrepreneurial in our endeavors. In the female-owned business market, Black women are starting companies at rates six times higher than the national average. As a category, businesses owned by Black women generated $44.9 billion in 2013.[8]

As Black women stare down structural challenges, we must also confront a culture that is often hostile and exclusionary.

To be a Black woman in America is to experience a constant barrage of micro-aggressions, some subtle, some not so subtle, in every sphere of American life. Micro-aggressions flow from a set of assumptions about Black women and their extraordinary attributes—our intelligence, our character, our fundamental worth as citizens—are constantly under assault in this great, complicated country.

I would know. I confronted micro-aggressions and the toxic assumptions motivating them that day in Charlottesville. I encountered them, daily, in the White House pressroom as one of the few Black female reporters doing the People's Work. Indeed, I have encountered them most days of my career. They are often snide and subtle—but always devastating. They are devastating because of what they imply about you—your talent, your professionalism, and, ultimately, your worth.

It's easy to laugh now about the preposterous treatment I faced during the last administration, but the experience was hurtful and, at times, traumatic. Press Secretary Spicer's constant insinuation that I was in the room only because I was a Black woman was painful.

When I asked a question about the Congressional Black Caucus, then President Trump asked, "Why don't you set up a meeting with them?" Days later, when I asked about Russia, I was told to "stop shaking your head."

As I did my job, administration officials insinuated that I was interested only in "personal fame." These attacks increased precisely when I offered legitimate criticism (never mind that the very function of an independent press corps is to criticize those with power; for four perilous years, we sorely tested that norm).

The sad truth is that while my experience as a reporter, who happens to be a Black woman, is extraordinary only because of where it happened—the People's House—it is far from unique. On a practical level, Black women navigate a combination of projected stereotypes, marginalizing treatment, and assumptions about our gender and sexuality every day. What's even sadder is that most people miss how we're treated altogether. Fundamentally, the Black female experience is one of second-class citizenship and yet, our experience remains invisible to most Americans. This is especially true of white people.

Americans do not sufficiently grasp or empathize with the ordinary, even monotonous, challenges faced daily by Black women across the United States. Even on the best day, it's uncomfortable, if not infuriating, to move through society as a Black woman. At work and even at home, we battle society's alternating perceptions: Sometimes we are "angry" and other times we are completely invisible. One moment, we are lusted after; in the next moment, our choice of dress, the way we keep our hair, or comport our bodies is criticized.

We are always expected to fulfill our duties, which, more often than not, serve others without any expectation of respect or acknowledgment—whether in the form of recognition or reward. For the Black woman, there's rarely a tribute or recognition of our lived experience. We can never just be. Instead, we relentlessly confront society's expectations of our role, our place.

Even the most powerful among us struggle. Consider Serena Williams, who is, by acclaim, the greatest tennis player in history and perhaps the greatest athlete the world has ever known. Despite her achievements, more than any other figure in all of sports, Williams endures a greater share of bad calls, poor treatment by the press, and unrelenting scrutiny of her body as "hypermasculine" and "scary."

And yet, when she was preparing to welcome her child into the world and she warned doctors of her family history of blood clots and said that she believed that she was experiencing a pulmonary embolism, the medical professionals tending to her care ignored her feelings and lived experience. In so many ways, they demanded her silence and, in that moment, the world's greatest athlete was invisible. She was experiencing a pulmonary embolism and biased, discriminatory decisions by her doctors led to a litany of painful complications. They almost cost Williams her life.

The social science affirms Serena Williams's experience. Black women's pain is far more likely dismissed by our medical system, often with perilous and tragic consequences. Our babies are born sicker and die at rates far higher than white newborns. In our education system, our aptitude is overlooked, even if we are policed in schools and disciplined in classrooms at rates far higher than our white counterparts. In the workplace, we experience more racism and sexism than any other segment of the American population. As our contributions are taken for granted, we are overlooked for promotions. Nevertheless, we engage in the world of work at rates far higher than white women—all while holding together our households, our communities, our country.

Yet all too often, we walk in shame and silence, believing the problem is us. However, the truth is that the establishment—from the medical industry to our politics—does not care enough about us to focus on us. The perception that Black women are strong and can take the pain and just handle it governs how we are treated. We are perpetually discounted.

I can personally attest to this. When I was pregnant with my first daughter, Ryan, I arrived at the hospital; I was filled with fluid and was scheduled for induced labor that quickly became an emergency C-section. Even though Ryan was my first delivery, I knew something was wrong.

I delivered my child and immediately after, I couldn't breathe. I panicked. I told everyone at the hospital who would listen, "I can't breathe."

They ignored me.

I was told, over and over and over, that "it will be okay" and "the baby must have pushed my diaphragm into my chest."

I continued to gasp for air while the hospital staff buzzed around me, alive with activity and unconcerned. It turns out that I was very sick.

Finally, a kind, attentive nurse who had come in to gently massage my back noticed that my breathing was off. She immediately called for a doctor.

Pneumonia had filled both my lungs with fluids and no one had caught it. The medical team was astounded. Once reality set in, the questions began: Why did this happen? Did I have a heart attack? Why wasn't my body pumping the fluid out like it should?

Even with that understanding, the medical team continued to ignore me. I will never forget the nurse who failed to answer my calls when I was in the critical care unit, or when I went for my chest scan and no one helped move me onto the table. Out of breath and panting for my life, I moved myself onto the table as my heart raced right out of my chest. Silently, I suffered. Big, fat teardrops streamed down my face. I was scared for my life. But I was more scared for the life of the little baby I had just brought into this world.

Here's an excerpt from "Black Newborns More Likely to Die When Looked After by White Doctors":

> Black newborn babies in the United States are more likely to survive childbirth if they are cared for by Black doctors, but three times more likely than white babies to die when looked after by white doctors.... The mortality rate of Black newborns in hospital shrunk by between 39% and 58% when Black physicians took charge of the birth, according to the research.... By contrast, the mortality rate for White babies was largely unaffected by the doctor's race.... While infant mortality rates have fallen in recent decades, Black children remain significantly more likely to die early than their White counterparts.[9]

Ryan is a strong, independent young woman today. Thank God. I am so proud of her. But I remain protective and vigilant. I am also inspired by social entrepreneurs like Ana Rodney, who runs a Baltimore-based organization called MOMCares, which is devoted to advocating for and serving Black mothers as they navigate American health care. I have such empathy for her work because I know the stress and the stakes confronting a Black new mother. From the moment my daughters took their very first breath on earth, I've had to take care to protect them, and me, from a system and society that weren't built for us. With a clear-eyed realism, I have raised my two daughters to understand that if we do not keep constant vigilance in our society, the consequences can quickly veer into the territory of life and death.

CHAPTER 5

Our Sacrifice

> *Recognize the trauma.*
>
> Fredrika Newton, president of the Huey P. Newton Foundation and his widow

PAIN, TRAUMA, AND OUR BODIES

As Black women strive in pursuit of our personal and professional goals, we must also navigate a culture that oversexualizes our bodies and dehumanizes us. Cultural touchstones, like the book *To Kill a Mockingbird* or the films *The Help* and *The Blind Side*, were ostensibly created to elevate Black narratives. Ultimately, however, they present one-dimensional portrayals of Black women to serve stories of redemption about white people to white people.

In an interview with *Vanity Fair*, actress Viola Davis said:

> Almost a better question is, have I ever done roles that I've regretted? I have, and *The Help* is on that list. But not in terms of the experience and the people involved because they were all great. The

friendships that I formed are ones that I'm going to have for the rest of my life. . . . I just felt that at the end of the day that it wasn't the voices of the maids that were heard. I know Aibileen. I know Minny [played by Octavia Spencer, who won a best-supporting-actress Oscar]. They're my grandma. They're my mom. And I know that if you do a movie where the whole premise is, I want to know what it feels like to work for white people and to bring up children in 1963, I want to hear how you really feel about it. I never heard that in the course of the movie.[1]

None of this is by accident.

When we look further back in our history, many of today's microaggressions are rooted in structures and systems that were maximally oppressive. Dating back to the very beginning of the transatlantic slave trade, our foremothers endured inconceivable burdens and tolerated slights, big and small, fundamentally rooted in the denial of their humanity. It is important to fully understand their experience and example, even if most of their names are lost to history. Individually and collectively, their lives are further proof that Black women in America have toiled for centuries without the gift of grace.

Our foremothers endured dehumanization on a global scale and their treatment laid the foundation for many of the interactions we are forced to tolerate today. For example, the hypersexualized trope of the Black woman as the jezebel extends all the way back to the African woman plucked from her tribe by colonizers who raped her with their eyes and their white penises. The stories of the atrocities faced by African women, captured and enslaved, are endless.

The stories from chattel slavery are particularly sickening when one learns exactly how women were chosen for the journey to become unpaid, shackled workers who, in almost every case, bred the next generation of enslaved, unpaid African men, women, and children cultivating this land and all her bounty.

A perky, firm bosom determined whether the African woman was to be free or enslaved. The captors determined their fertility and strength in part by looking at their breasts. The would-be slave was left behind in the motherland if her breasts sagged, as she was already a mother and had breastfed. She was simply tossed because, as the theory goes, she was spoiled for further breeding.

The primitive judgment of a woman, an African and Black woman specifically, set off centuries of degradation and standard setting endured by Black women in the United States and around the world. That horrific treatment supports Shirley Chisholm's sentiment that being Black and a woman is a "double whammy."

Of course, it didn't stop there. Those women were stripped of their family and tribe, and any protection they might have enjoyed. Alone, terrorized, and dehumanized, Black women had to fend for themselves. On plantations and throughout cotton fields, Black women stood alone for themselves, their families, and their fragile communities. Black men, beaten and broken at the slightest hint of insubordination, could not fight for their honor as families were broken apart and worse.

IMPOSTER SYNDROME (OR WHY WE DON'T CELEBRATE OURSELVES)

When it comes to the truth about our Black women who were enslaved in this nation, there are some very hard truths we have not digested, even as the facts bare it all. Just ask Valerie Jarrett, the former senior advisor to Barack Obama.

"I have a bill of sale for my great-great-grandmother, [including] who she and her children were sold to. A bill of sale, just like a piece of chattel. How could that be? And yet, we still have an infinite capacity for joy and to love and to be generous of spirit and to show grace and forgiveness. That's something else that's a strength of character and so that they could not take away from us."

Professor Henry Louis "Skip" Gates, of Harvard University, conducted his famous genealogy study on Jarrett's family:

> He [Professor Gates] did this whole genealogy and I mean, just think about it: My great-grandfather was the first African American to go to MIT in 1890. His father's name was Robert "Henry" Rochon Taylor. Henry Taylor was born a slave and he was freed after the Civil War and went on to become a carpenter. He saved enough money to send his son to MIT. Henry's father was the slave master. And, so, my great-grandfather's father was born from the rape of his mom. And where is she? She's invisible. Because she was [written] down as a slave.

That particular experience of invisibility, then and now, contributes to what Jarrett relayed as the "Imposter Syndrome." For Jarrett and many other Black women in leadership, decades of messages that convey that you do not belong are followed by feelings of not deserving to belong once you've "made it." Those feelings are hard, if not impossible, to shake, following women into even the most distinguished careers.

For many Black women, however, feelings of the Imposter Syndrome obscure a deeper anger, or, dare I say it, rage that we repress. It is well documented in literature and across the culture the extent to which Black people, especially Black women, have had to "choke" down their rage in order to conform to a society that paradoxically fails to invite their involvement. Black women have to pretend not to notice the slights, the stereotypes—the struggle that characterizes so much of our experience in this country—and it is enraging.

For all these reasons and so many more, Black women are not given grace, unless we give it to ourselves.

Ever since Black women arrived in this country, the ever-present and overwhelming trauma that Black women carry and experience has never been fully acknowledged. We are broken—generationally, collectively, and personally; however, we have learned to move through that trauma in our struggle for survival, with only the grace we afford ourselves.

The extent to which Black women have endured trauma is shown in our struggle to rise above it and the invisibility society ascribes to us as we do so.

Fredrika Newton, the widow of Black Panther Party cofounder Huey P. Newton, calls out the trauma of fighting for survival against oppression. What the Black Panther Party did as a militant voice, and actor, for racial justice left scars even as Fredrika Newton says the movement was all about the love of and for the community. The assassination of Fred Hampton by members of law enforcement was deeply traumatic, and impossible to forget or forgive. He was an incredibly important leader and voice for a unique, revolutionary movement. In the years after Hampton's murder, when Huey P. Newton was involved in an altercation that left an Oakland police officer dead, the chasm between law enforcement and the Black community only grew wider. The trauma from those years is undeniable.

"I'm definitely talking about trauma and recognize it too. And you have been traumatized. Are you kidding me? You've been on those front lines. Just don't minimize what you do and the impact on you either. So the front line doesn't always mean on those streets. The way that this thing hits you is on so many levels. So recognize the trauma."

The trauma is real.

Fredrika Newton walked into her power as a Black woman much earlier than I did. She was a teenager when she first fought against the structural injustices confronting Black people.

For me, my ascension into Black womanhood arrived the year I turned forty. It was not a celebration at all. In other cultures, the community comes together to celebrate when a child elevates into manhood or womanhood. It is joyful. For me, that was my worst year. It was the year I lost my shero, my mother—the woman who led our family and delicately slayed dragons like it was no big deal. My mother left this earth, leaving me as the woman in charge. It is the Circle of Life, and I was an unwilling participant in this natural tradition.

There are no books to describe what happens next; there are no words to guide you when you find yourself alone to shoulder the burdens of family and community and career—hell, the world. When you enter into the sacred society of Black women "dragon slayers" or of those who know how to get out of a fix, you lean on the experiences of other Black mothers whose lives serve as a guide. Like a good law professor, my mother would rely on past precedent as examples of how to navigate the Black womanhood journey. She embodied the celebration of *Sankofa*: She often spoke of this African practice of flying forward while looking at the past. Her life was the lesson.

Black womanhood is one of those monumental equations in this nation, where the formula is dissected to re-create the magic, but it has never been replicated in any other community. Face it, women are amazing and forge ahead. However, for Black women, the tenacity, raw willpower, and force to sustain themselves in those crucible moments created this resolve and continual push to be better than the best. There is no rule book for this unique yet communal experience. Each stride of her story chronicles the collective understanding of who we are and how we do it against seemingly insurmountable odds.

The commonality is strikingly present. As in Africa, the sisterhood for Black women in America is a required unit of support and fortification. There is strength in the circle of commonality and yet unique difference. Black women in America are close and yet so far from our sister lineage in Africa that we cling to it in many ways, trying so hard to understand. Especially in skill, wit, and strength, Black women in Africa marvel at us and we marvel at them. They carry the household, children, husbands, and community all in that basket over their hearts and American women carry it all over in their very being.

I will never forget greeting my mother on more occasions than I could count as my sister-friend. She taught me that Black women elevate. However, more often than not, those new heights come after great loss or struggle.

Like my mother's, Fredrika Newton's life is an exercise in wisdom. To this day, I look to her for lessons in how to navigate today's struggle. Newton says healing is a must:

> Recognize the impact that mentors have—they [are a kind of] spiritual program, whatever it is, and really take good care because it's a wonder that we're not all screaming, running crazy

down the street. It really is. So I think that I say it's vital to keep young energy around you. At least I do. I have to. There's no way I could continue to do the work that I do without having these young minds. They're so agile. They're so brilliant. They're so courageous. They're so innovative. They're so great, you know, so I keep young people around me to be able to continue to do the work that I do. And I have to be balanced. Otherwise, I'm at work. And I'm sure you are, too, because there's so much work to be done. But what we're in here for, it's a marathon. We're not sprinting.

Through the generations, Black women support each other through the juxtaposition of grief and celebration as we dance, every day, with the inconvenient truths that mark our life's journey. Our truth is that small moves forward, little moments of laughter, and sisterhood represent the building blocks of progress. We all have pain. We all have trauma. We love each other to survive and move forward.

The larger celebration of Black women and our contributions has long been absent from the conversation.

I will never forget talking to Bishop T. D. Jakes, who was very complimentary about my books. He emphasized how I kept winning against a moment in history. The bishop implied that I was winning against Donald Trump. He asked if I ever celebrated myself—or my success as a journalist.

I was shocked that he had asked.

I should not have been. But it is not surprising. As a matter of practice, we Black women place ourselves behind the issues of the day—whether they are the issues at home, in the church, or across the community. I told the bishop, "No, I don't celebrate myself." But the exchange marked a turning point for me. Today, I proudly celebrate myself and the contributions of other Black women every chance I get. Black womanhood is everything. The sisterhood is strong.

There is no weakness in knowing when to fight and when to rest to reassess the lay of the land and those we are fighting against.

COVID-19 was my attempt to keep my family safe, but I also used it as a time to breathe. I worked from home for a year, uncertain about what was happening. I hid from physical threats of violence that spewed from the far right. And I hid from the deadly invisible germ that Trump did not care about.

I am that person who has multiple comorbidities. I chose to stay alive and in doing so, I was able to regain the strength to fight more. If you can't take care of home and fight for it, then why be on the battlefield at all?

My heart is home with my kids.

TAKEN FOR GRANTED: SERVING OTHERS FIRST

Black women always put others first, even if it means putting ourselves last.

For centuries, we have sacrificed our bodies, our hearts, and our ambition for the common cause. Sacrifice and its ugly twin, suffering, are core to our experience. Tragically, sacrifice and suffering are part of our identity. They are embedded in the stories we tell about ourselves. A detailed and empirical study of literature found that the narratives about Black women revolve primarily around our "sacrifice, abuse, and injustice."[2]

When we talk about Black women or even when Black women

talk about ourselves, there is shockingly little adventure, romance, and happiness in our story—there is so little joy.

Why?

Well, for starters, narrative is shaped by fact. And the facts are as stark as they are clear: the sacrifice Black women experience is structural.

The deeply traumatic experiences of the transatlantic slave trade, chattel slavery, and centuries of struggle for freedom and equality have left a deep imprimatur on our psyche. Every episode of the last four hundred years has demanded unrivaled sacrifice from Black women. Black women have endured the disintegration of family and the assault on our bodies. We've internalized the struggle, the responsibility, and the burden of keeping our families, our communities, and ourselves together while everything else fell apart.

We can still feel their pain today. For example, enslaved people who made the brave and courageous choice to marry, despite the likelihood of separation, amended their vows: "Do you take this woman or this man to be your spouse—until death or distance do you part?" Black women were forced to accept the inevitability of their children being shipped to another plantation or another state, never to be seen again. Black women endured assaults on their bodies. Nothing was sacred.

Perpetual sacrifice forced us to develop superpowers. Literally. In a landmark study, Amani Allen, PhD, captured the unusual ways Black women cope with a racist and sexist society—what she calls our "superwoman schema." Those powers include "feeling an obligation to present an image of strength, feeling an obligation to suppress emotions, resistance to being vulnerable, a drive to succeed despite limited resources, and feeling an obligation to help others."[3]

Remarkably, Allen demonstrated that when Black women experience racial discrimination, their stress response lowers. Counterintuitively, when Black women experience racist attacks, they transform feelings of rage and anger and frustration—emotions highly correlated with negative health outcomes—into emotional resilience. Allen states, "Although contradictory to studies showing that emotion suppression is bad for health, this finding makes sense, in light of research showing that the most common emotional reaction to discrimination is anger, and anger is bad for health."

Allen's study contributes to a growing body of research demonstrating how the stress associated with racial discrimination becomes biologically embedded. The burdens of racism and sexism have forced Black women to temper their biological response to the racist and sexist society we live in. Allen reminds us that "the problem is not that Black women need to learn to cope better with racial discrimination. The problem is racial discrimination itself and the need for interventions intended to address racial discrimination as experienced in the workplace, by police, and in society at large."[4]

For better or worse, and unsurprisingly, Black women are leading the intervention. There is no better example than Super-Vote-Getter Stacey Abrams, who has spent the better part of a decade laying the groundwork for historic political victories in Georgia. Despite heartbreaking loss, years of sacrifice, and great personal expense, Abrams delivered not only for America but also for Democracy.

"Black women tend to be the targets or the victims of the most incendiary and heartless policies," says Abrams, who serves as the founder of Fair Fight, which is dedicated to addressing voter suppression. "But our resilience ensures that we not only persevere but we tend to raise awareness and create the platforms and projects by which we push back and survive."

Let us go back to that fateful night in 2018, after Abrams's crooked loss to the now Governor Brian Kemp. In her "concession" speech, Abrams said it plainly for all to hear: "We are a mighty nation because we embedded in our national experiment the chance to fix what is broken." She vowed that night to fix Democracy in Georgia. And she did just that. Always "the Fighter," Abrams generated progress from her pain. In an election cycle fixated on Midwest working-class white voters, Abrams understood that Black women are the Democrat Party's true "Blue Wall"—the progressive movement's most vital voting bloc and its essential leaders and activists. When Trayvon Martin, Eric Garner, and Philando Castile lost their lives at the hands of law enforcement, it was Black women—three of them to be precise—who founded #BlackLivesMatter.

For decades, Abrams has done the work in Georgia. And she has done it in plain sight. Imagine her surprise at the media's surprise in November 2020. Recall the fawning coverage that followed when Abrams delivered the presidency and then the Senate to Joe Biden. Reporters exalted in the fact that she spent years of her life turning Georgia blue. But what was missed was the personal cost. What was glossed over was the perpetual discounting and dismissal of Abrams's talent. Rather than questioning those facts, the media glorified her sacrifice by elevating her suffering to the level of expectation.

"Well, of course, Stacey Abrams did this. It's what Black women do!" Stacey Abrams's response is on point:

> The personal dismissal of me [goes back to what] I was saying would happen. It did not seem likely that a Black woman would be the person to transform the electorate of Georgia. If anything, if a Black person were going to do it, they presumed it would be a

Black man because in American history, Black men were the only ones to be governors. They were the first to be senators. The Black men have broken through before Black women. To the extent there was any countenance of my ambition for our work, there was an assumption that it wasn't going to be a Black woman who did it.

I would describe it less as "shock and awe" but rather "ignore and surprise." I said it many times over the last decade. I didn't give the full brushstroke of what I thought would happen, but my conversations about Georgia and the importance of fighting against voter suppression have been a constant narrative. The difference is that until 2018, no one paid attention, and until 2020, no one believed. I didn't spend my time in public luster. I spent my time in quiet building, working to create an organization, to bolster other organizations, to craft strategies, to attack, and to anticipate other broadsides that were going to come. It was a quiet process because there was so much to do.

Sadly, this isn't anything new. "Black women have historically been a central part of protest movements and civil rights action throughout our history," says Abrams. "Our efforts are typically the spine of these movements, but I also recognize that we are rarely the initial beneficiaries of success." Abrams understands that it's determined, tireless, and intentional work on the ground that wins elections. And so much of her power is tapping into the passion and commitment of Black women across Georgia. And not just professional activists and advocates but everyday citizens who want to help Georgia get it right. Extraordinary Black women like Louise Terry, of Putney, Georgia, a small farm community of just a few thousand people. Louise is a quiet trailblazer who marched with Martin Luther King Jr. Everyone's auntie, nearing eighty years of age, and inspired by Abrams, Louise knocked on almost every door in Putney in the "long shot" effort to deliver Georgia for Biden. Super-Vote-Getter Stacey Abrams's secret weapon? People like Aunt Lou, of course.

Abrams is, of course, a wise student of history. She takes her cues from past fights. And history shows us the persistent power of Black women in social movements. After enslavement, for instance, Black women led the fight for equality, sacrificing ourselves for the "larger" cause. First, Black women stood with Black men in their fight for The Vote immediately after the Civil War. That fight was complicated by the fact that Black women also fought in the Civil War and contributed mightily to the twin causes of abolition of slavery and Union victory. To illustrate the point, we need not look further than Harriet Tubman, who commanded combat troops during the Civil War, served valiantly as a spy for the Union, delivering vital intelligence, and guided enslaved people to freedom via the Underground Railroad.

Second, Black women stood with women everywhere, including white women, for suffrage even as protections against race-based discrimination eroded over the course of Jim Crow and well into the twentieth century. This point is important. American slavery was unique in its racist foundation. The discrimination and racist ideology that propped up the institution of slavery did not vanish when it was abolished. Black women endured America's horribly racist caste system that materially deprived them of all rights, even while fighting for the enfranchisement of women everywhere.

Indeed, our sacrifice is structural.

The expectation of the strong Black woman is often weaponized as an excuse by others—our romantic partners, our work colleagues, our friends, community members, service providers—to rationalize our poor treatment. Our medical workers can overlook our pain and ignore our complaints of poor treatment because our bodies can supposedly "take it." We can carry more of the water at work for less credit because, subliminally, so many of the people around us believe it is our role—to sacrifice, to give on behalf of the larger cause, to suffer, to endure.

"[People discount)] the effectiveness of Black women as organizing masters," says Abrams. "[We are] people who can navigate these difficult and complicated systems to deliver change."

Abrams is right. We are taken for granted.

I personally experienced an incredible case of being taken for granted—indeed, dismissed—in the lead-up to the NAACP presidential candidates' debate in 2019. The candidates were Democrats and Republicans. Invitations were sent to the entire field in both parties, including Republicans William Weld and President Trump.

Before the event, I had the opportunity to ask the president in the Oval Office if he was participating. He asked if I was moderating. I responded that yes, I was. And from the Resolute desk, Trump said he would not take part in that Detroit debate.

Fair enough.

Meanwhile, the NAACP continued to negotiate with both parties to determine who would participate in the debate. I was still set to moderate. Behind the scenes, I was being stabbed in the back. A quasi-journalist (and talking head), who happens to be a Black man, turned on our community and started negotiating to have Trump participate in the debate, with the provision that he would moderate the debate instead. This individual weaponized my track record as a journalist and my fight for our community, our priorities, and our cause—a track record that invited death threats and caused my hair to fall out—to ingratiate himself with the president. He knew that the odds that Trump would appear would naturally increase if I were not in the picture.

Cold.

I was made aware of this person's maneuverings, but I kept it moving. That's what we do. Remember, when the pressure is on, our stress response lowers. In the end, the NAACP moved forward with me as the moderator of the event. In the process, another journalist revealed himself as Trump's pawn.

STANDING UP, FINDING JOY

Former Atlanta Mayor Keisha Lance Bottoms rejects the notion that we have to accept being underestimated. In our interview, Bottoms admitted to being underestimated and discounted by people from all walks of life, including Black men. Her experience resonated with mine and reminded me of the shenanigans around the debate when that other journalist, and a Black man at that, nearly succeeded in having me removed as the moderator of the presidential candidates' debate. He was close enough to create a stir. Ultimately, he did not hurt me with his underhanded tactics, which would have brought Trump into the discussion had he succeeded.

I have learned from Bottoms's example and have been inspired by her grace. When other members of the press corps steal credit for my stories and scoops, I channel Bottoms. I hold my head high and walk with grace. I reject that standard of treatment. I make it known—respectfully, of course, and always—that I won't tolerate disrespect in the workplace or beyond. I also keep my professional circle tight—integrity is the price of admission if you are going to work with me. After all, when you know better, you do better.

I've learned from Abrams and from Kamala Harris. In their own ways, each of these women stands up for what is right and each one stands up for themselves. They know their value. They subvert the cultural expectation of sacrifice that has literally rewired our bodies and stripped our spirit of the joy and wonder to which we are entitled.

I love Stacey Abrams's most radical act of liberation: Selena Montgomery. Our heroine, the Super-Vote-Getter extraordinaire, leads a double life as an author of fiction under the nom de plume of "Selena Montgomery." How does a romance novel liberate, you ask? Her books are full of romance, adventure, and joy. So much for sacrifice, abuse, and injustice. Stacey Abrams proves that the Black woman can save the world and save herself. She proves that amid the generational struggle, we can feel and celebrate ourselves, and live out our joy. She rejects the premise that Black women are simply social justice machines—unfeeling, selfless automatons who turn out voters and make it home in time to, well, do everything at home. No. We deserve love. We deserve respect.

We deserve to experience our full humanity.

CHAPTER 6

Our Voice

> *He's a liar! He's a con man. He's a cheat.
> And we've got to stop his ass!*
>
> — Representative Maxine Waters

FIGHTING FOR OUR SEAT, MAKING OUR TABLE

I have learned so much from Stacey Abrams.

As a writer of fiction, "Selena Montgomery" has taught me the importance of nourishing the sources of joy in your life in order to carve out a little space for you. This is so important because most Black women lead lives that put others first. From a young age, we internalize the notion that it's never about us. That can lead you to a place where you not only deny yourself grace, but you can also deny your own voice. As the warrior at the helm of Fair Fight, Abrams shows us too that it's important to elevate your voice, to use it for all and for whoever is worthy of it. This is all the more important in a society that relentlessly tries to silence and marginalize your voice. Abrams's example gives lift to Our Voice.

Watching Abrams, it's clear as she pushes back against her opponents that she understands the value of her voice. Her struggle is rooted in the fight for fairness. She speaks up precisely because others cannot. Lo, my dear reader, beware the Black woman motivated by a righteous cause. In her fight, Abrams has achieved so much: She flipped not one but two US Senate seats and helped deliver the White House to Joe Biden. Most important, she fired up our imaginations by pushing us to envision what's possible in and for our country if we organize in those communities that we underestimate or overlook altogether.

Abrams's work at Fair Fight begins and ends with her commitment to voting rights. Her 2018 gubernatorial bid underscored the importance of voting rights. In that race, the Georgia GOP perverted and distorted the electoral process. The GOP disenfranchised Black voters and suppressed Black votes. These are ugly American traditions and a normal occurrence in Abrams's life in the South. When she was *this close* to breaking one of the toughest glass ceilings in American history, those ugly traditions reared their heads again.

It's telling that Abrams, instead of pursuing another political race or a federal office, doubled down on the fight for voting rights. She is one of America's most principled architects of progress, and her priorities are a barometer of what's important. She is strategic and understands what matters most for the American people.

In America, The Vote is paramount. It's the yardstick by which we measure participation in the American Experiment. It's our Republic's most essential tool for which countless people have fought and died since the Revolution.

The Vote is also not a given. At the nation's founding, The Vote was restricted to white men with property—just 6 percent of the population. By the 1820s, white men won universal suffrage after organizing and agitating at the state level. As white men consolidated political power in the United States, other groups, beginning with Black men and extending to women of all races, confronted a backlash—a pattern that has continued to the present day. That backlash was part of a larger effort to systematically oppress anyone who wasn't male and "white," a category that has expanded through time.

Black people were property while white women were mere extensions of their husbands or fathers. Black women were invisible. Their vote wasn't even on the table.

But it was on the horizon. The movements for abolition and women's suffrage, organized for generations, fought for the basic humanity of all people in the United States. Those sheroes and heroes laid the groundwork of the modern movement for voting rights—so that we could participate, so we could begin to shape the country in our image. It was only after the abolition of slavery in 1865, after the victory of women's suffrage in 1920, after the toppling of Jim Crow in 1964, and after the enactment of the Voting Rights Act in 1965 that Black women, finally, had a formal, real, and powerful seat at the table.

That seat was made available to us just over fifty-five years ago, in 1965.

Unsurprisingly, my parents raised me on the importance of voting. It was in our bones. Even as a small child, I could sense that our exclusion from America was so very recent and our inclusion was so very new, and fragile, and insecure. It could be lost at any point.

I grew up internalizing a sense of responsibility. My generation was entrusted with something sacred and precious. I have earnestly, and with all my soul, passed this core value down to my daughters. My girls are (only) occasionally annoyed by how passionately I have emphasized The Vote in their upbringing. Their annoyance is just adolescence. They too understand: Voting is our birthright. We must defend it, at all costs.

It's natural that speaking up on voting rights is core to my reporting. Perhaps more than most, I understand its importance—historically, socially, and politically. In recent years, the stakes around voting have only increased. In 2018, when I asked the president of the United States a question about voting rights after widespread reporting and recognition of voter suppression, I was doing my duty—as a journalist and as a Black woman in journalism.

Not for the last time, the administration tried to silence me.

That same day, the press initially zeroed in on what happened to CNN journalist Jim Acosta, who had his press credentials rescinded at the same press conference: "You are a rude, terrible person. You shouldn't be working for CNN," President Trump said after Acosta, following a long and arduous exchange, asked him a question about the Russia investigation.

Judging Acosta's experience and my own, what has since become clear is that President Trump had a pattern of silencing those with whom he disagreed or, in my case, those of whom he disapproved—anyone different from him. When I asked a question about voting rights, the Leader of the Free World told me to "sit down" before concluding that I was "hostile." In just a few curt words, the president attempted to take away my voice—and negate my lived history. I wish I could say that this was a unique incident, but that interaction played out over the course of that president's administration with both him and his parade of press secretaries.

Of course, I did all I could to counter the president. Despite the hostility of the administration and the risks to my career, I elevated my voice. Like Abrams, I understand the stakes associated with my work. If I don't ask the question, then the question will not be asked. After all, the number of Black women operating in the White House press corps is tragically small. I am raising and reframing the issue of silencing Black women now because Black women's voices have been silenced throughout history. No more.

We grapple with the truth and confront our historical record as a country.

Black women have almost always seen their voices discounted during our history's most pivotal moments: the fight for abolition of slavery, the cause of women's suffrage, the civil rights movement, the Vietnam War, the wars in Afghanistan and Iraq, and throughout the fight for our voting rights.

Singer and actress Eartha Kitt was blacklisted and tormented by the CIA after she spoke out against the Vietnam War and its connection to juvenile crime in the United States at a luncheon with President Lyndon B. Johnson and the First Lady Bird Johnson in 1968.

Entertainer Josephine Baker renounced her US citizenship and was put on an FBI watchlist for her activism and efforts to desegregate the country.

Activist Angela Davis earned her notorious spot on the FBI's "Most Wanted" list for her alleged ties to communism and her direct involvement with the Black Panther Party. Davis was smeared and discounted because her voice was perceived as dangerous. They—our government—wanted her silenced because she was powerful.

As we became more powerful, the government's efforts to shut us down were less successful. My personal shero, Congresswoman Shirley Chisholm, refused to stay silent about her beliefs, including and especially her disdain for the tragic Vietnam War, despite the backlash her positions invited. She was truly "unbought and unbossed" (per her autobiography). Chisholm stood by her truth during her presidential campaign, even as she experienced sexism from within the Black community when members of the Congressional Black Caucus voted for her white counterpart, George McGovern.

Brother Malcolm X had it right when he said, "The most disrespected person in America is the Black woman."

Perhaps more than any other demographic group in the United States, Black women confront almost daily a hostile and pervasive culture that works to silence us. A litany of micro-aggressions, racism, sexism, and dehumanization show up in our politics, at work, at the doctor, in the classroom, and even in our own communities. For Black women, that culture of silence is consistent with what scholar Moya Bailey coined as "misogynoir" to describe the misogyny directed toward Black women, whereby race and gender both play roles in the biases directed toward us. In other words, we are ignored for being female *and* Black.

Few people living have endured as much "misogynoir" as Representative Maxine Waters.

In the days leading up to January 6, 2021, now infamous for the white supremacist insurrection that was organized against the United States Capitol, Waters—"Auntie Maxine"—sounded the alarm.

A white supremacist had sent Waters her second death threat that week. Waters could sense that this was just the beginning: She warned the House speaker and the Capitol's sergeant at arms of the impending danger. President Trump just might try to overthrow the election at the Capitol on January 6. They assured her that everything was under control. She made the query in the Democratic caucus meeting and in the following days, the Capitol police gave members and staffers a memo discussing how to protect themselves if these protests took a turn for the worse. When it came to the expectations of Capitol Hill policing authorities, the failure was deadly.

They were very, very wrong.

For Congress's longest-tenured Black woman, it was a familiar position. For decades, Maxine Waters has served America as "The Prophet." Early on, she called for divestment from South Africa's apartheid regime; Waters was among the first to oppose the Iraq War. No matter the president or the party in power, Waters has always been a truth teller and a seer, capable of peering into the future to warn us all of coming danger. Waters called for the impeachment of the recently inaugurated President Trump as his conduct quickly escalated to the level of "high crimes and misdemeanors."

"He's a liar! He's a con man. He's a cheat. And we've got to stop his ass!" she exclaimed in a speech in May 2017. Her comments drew instant ire as did my question in the White House Briefing Room around the same time, when I challenged press secretary Sean Spicer on the administration's perpetual state of crisis, two-and-a-half months into the president's term.

The condescending blowback from powerful white men like political commentator Bill O'Reilly and Spicer generated a larger national response from Black women throughout the nation. They spoke up—many for the first time—about the mistreatment, bias, and discrimination they face just for doing their jobs. #BlackWomenAtWork galvanized Black women everywhere and drowned out the thousands of Twitter trolls and white supremacists who sent threatening emails to the congresswoman and me. We forged a special bond as two Black women unafraid to fight for what is right at the highest levels of our democracy. In her first speech after her painful and unexpected presidential election loss, Hillary Clinton made a point of mentioning

Maxine Waters and me in her remarks, highlighting the racism and misogyny that we as Black women faced for just doing our jobs.

DOUBLE STANDARDS

The culture of silence targeting Black women is just that—it is a culture. It's pervasive.

"Were you silent or silenced?" Oprah Winfrey asked Meghan Markle in their bombshell interview about Markle's experience of being suppressed by the British Royals. Markle was dismissed and ignored despite her cries for help when she was having suicidal ideations. They stifled her voice when she tried to stand up to the press. They ignored the racism she and her then-unborn baby endured.

Throughout that pain and beyond, I hope Markle finds support in sisterhood. During the nightmare of the last administration, I found sisterhood in so many of my relationships with Black women. Miraculous sisters like Maxine Waters, who, like me, endured countless death threats across the life of the administration. To have someone in whom I could confide the trauma of that entire ordeal was priceless. When your hair falls out because of death threats from racists, you will need someone to talk to who can relate!

Relating through pain is familiar territory for Black women.

OUR SILENCE

Black women are unstoppable forces. As truth tellers and as agents of change with a voice, we speak our truth for the betterment of our country, even as society refuses us a platform and doesn't heed our warnings.

In order to be heard, Black women have to find ways to subvert the system.

When Black women are silenced, we find creative ways (such as social media) to speak our truth. Even if a Black woman chooses to be silent, that's strategic. For instance, Black women have a particularly complicated relationship with their hair. In society, our hair is either fetishized ("Can I touch it?") or demonized (see the countless laws, regulations, and decisions by leaders and systems to expel Black students and workers from communal spaces because of their hair).

Just as Ayanna Pressley had finally embraced her Senegalese twists as her signature look and just as the press and her colleagues accepted them too, she suddenly found herself dealing with alopecia (a medical condition that causes sudden hair loss). On the night before the House voted on articles of impeachment against President Trump (for the first time), Pressley lost her last bit of hair. Rather than hide this intimate experience in her life, Pressley shared it publicly to make space for others who could relate. She elevated her voice through expression and fought back against the culture of silence that would have seized the opportunity to erase her presence in the Squad because of her hair. She said, "It's about self-agency, it's about power, it's about acceptance."

Or take tennis phenom Naomi Osaka who recently declined all press conferences out of respect for her own mental health and to raise awareness about anxiety and depression more broadly.

Unsurprisingly, Osaka encountered tremendous pushback on the decision. That didn't stop her. She took to Instagram and shared her truth. In fact, she even doubled down and withdrew from the French

Open to preserve her mental health. Osaka knew she didn't have the bandwidth to make a public appearance, so she used her writing voice instead. As a result, she received support from fellow Black athletes and other Black women across the world who are tired. The next generation is strong in its convictions and takes its cues from our elders, leaders like Maxine Waters. With our combined leadership, the tide might (slowly) be turning.

Osaka is pulling the curtain back on the routine exploitation of Black women in athletics and entertainment, and how easily Black women are taken for granted in these spaces. She also shows us how the next generation is doing things differently. Unlike her predecessors, who likely would have suffered through the experience silently, Osaka used her platform and her power to defy expectations that might entertain others but fundamentally detract from health and well-being.

Our next section explores precisely how the next generation is changing the game to not just save the world but to save ourselves.

Section III

The Promise

WHAT'S NEXT FOR US AND AMERICA

CHAPTER 7

"A Little Child Shall Lead Them"

> *Sometimes we choose our problems and sometimes our problems choose us.*
> — Tiffany Loftin, national director, NAACP Youth & College Division

BLACK GIRLS AND THE PAIN OF A CHILDHOOD INTERRUPTED

We often wait, no, search, for rock stars to save the day.

We forget that even rock stars were, once upon a time, regular people. With the fight against injustice, many of them started young in their fight for progress. After all, the Rev. Dr. Martin Luther King Jr. was just twenty-six years old when the Montgomery, Alabama, bus boycott started.

Heroes are not always world-shaping figures like Dr. King. More often than not, they're ordinary people. And you don't have to look far to find a hero. She is almost certainly our sister, a daughter, a niece, a granddaughter.

They are our young people and they're changing the world right here, right now.

One of my favorite passages in the Bible is Isaiah 11:6:

"The wolf shall dwell with the lamb, and the leopard shall lie down with the young goat, and the calf and the lion and the fattened calf together; and a little child shall lead them."[1]

I love that: "and a little child shall lead them." All those years ago, our ancestors envisioned a world restored and healed by peace—a world forever shaped and led by our young people. In that world, even the fiercest, most voracious foe is tamed by our children's essential goodness: Their sense of peace would calm the conflict that defines so much of our lives. Their purpose for a better world would realize a safe, secure, and fair future for all of us.

The youth often have the answer but only if we're ready to watch and listen. Of all the Black women who struggle with the invisibility and marginalization that accompany their identity, it is our young Black women who are overlooked the most. We do this at our peril. If history is any guide, our young people are driving transformative change today, right now, and inevitably will grow up to shape our world for the good.

Black girls are driving transformative change in society despite society. Like Black women, Black girls endure a disproportionate share of hardship as they pursue their education and early work opportunities. And just like Black women, Black girls become adept, at an early age, at converting pain into progress. The training of our supersheroes begins early. I wish it weren't so.

Our society is only beginning to pay attention to the struggle of Black girls. In a piece published by the *New York Times*, "'A Battle for the Souls of Black Girls,'" superstar reporter Erica Green (another exemplary Black woman in journalism), and her colleagues, Mark Walker and Eliza Shapiro, explore how Black girls are perhaps the most at-risk student group in the United States.[2]

For starters, Black girls endure disparate treatment in the rate at which they're disciplined in the school system as compared to white students. While Black boys lead all categories in the rate at which they're disciplined, Black girls are not far behind—but their struggle receives a fraction of the attention. The Common Application, which applicants may use to apply to nearly a thousand colleges and universities, recently cited the disproportionate discipline of Black girls as the reason for the administrator's decision to stop asking students to report whether they had been subject to disciplinary action. This issue is at the heart of the lawsuit brought by the NAACP Legal Defense and Educational Fund, led by the dynamic duo Sherrilyn Ifill and Janai Nelson (Ifill's successor at the venerated institution).

The lawsuit highlights a staggering reality: in New York City, Black girls in elementary and middle school are eleven times more likely to be suspended than their white peers, and Black girls are nine times more likely to be arrested. As Monique W. Morris, the executive director of Grantmakers for Girls of Color and author of the book *Pushout: The Criminalization of Black Girls in Schools,* said to the *New York Times,* said "We are in a battle for the souls of Black girls."

The reason for this outrageous treatment flows from a painful discriminatory phenomenon, the "adultification bias." Survey after survey indicate that Americans perceive Black girls as needing less nurturing, less protection, less support, less comforting; they are presumed to be more independent, and know more about adult topics and sex. In the study "Girlhood Interrupted," researchers highlighted the far-reaching and sweeping consequences of this harmful and systemic implicit bias.

Black girls are not only punished at far higher rates, but they also are cultivated and developed at far lower rates for leadership and mentorship opportunities. When people perceive you as "grown" well before adulthood, they invest less in you and hold you to higher, harsher standards. This is what researchers refer to as a "double bind." Just like Black women who often walk a tightrope of competing expectations, Black girls, from an early age, must navigate a world that treats them as different from all the other girls.

If there is one takeaway from this book, it's a call to action to treat our Black girls as who they are: girls. They deserve patience, cultivation, care, and support like any other girl and yet the research is clear: Black girls are subject to a painful adultification bias that deprives them of investment and makes them far more likely to end up disciplined or even incarcerated. Interrupting Black girlhood short-circuits the development of Black girls and puts them on a life path less healthy and far less supported.

This fact should reframe the extraordinary accomplishments of Black girls everywhere. They are leading, contributing, and changing the world for the better, despite societal forces. Imagine their power if we invested in Black girls instead of harmed them, if we believed in Black girls instead of discounting them.

An excerpt from "Simone Biles, Sha'Carri Richardson, and How the Olympics Failed Black Women" drives the point home:

> **The games have consistently made headlines for the wrong reasons, particularly for the mistreatment and discrimination of Black women athletes. From Sha'Carri Richardson's pre-Olympic**

suspension for smoking legal marijuana to the International Swimming Federation's ban on swim caps designed for natural Black hair, Black women are the common denominator in many of these stories. These rigid stipulations speak to the rampant racism and misogyny within the athletic world that directly impacts how Black women perform on the playing field—or if they even have a spot on the field at all.[3]

Black girls, particularly in the world of athletics, are using their formidable platforms to demand that we pay attention to their disparate treatment. The adultification bias leads too many to take for granted the performance of Black girls and young women, as if we are entitled to their performance for the benefit of our entertainment. Meanwhile, they are subject to harsher treatment—by the press, the regulatory bodies, and even the fans.

We should commend young leaders like Naomi Osaka, Simone Biles, and Sha'Carri Richardson for interrupting these toxic cycles and drawing attention to the rampant mistreatment of Black girls everywhere. We should also recognize that Black girls do not only endure this treatment, they also suffer under it. Indeed, there is a cost. The mental health crisis confronting Black girls is very real and incredibly concerning.

An excerpt from "The Case for Focusing on Black Girls' Mental Health" shows us the exploding crisis quietly unfolding among Black girls:

> Black girls have serious unmet mental health needs—it has been called a mental health crisis "hiding in plain sight." Between 1991 and 2017, suicide attempts by Black youth increased, while suicide

> attempts among youth across other race and ethnicities decreased. Actual suicide death rates for Black girls ages 13 to 19 increased by 182% from 2001 to 2017. . . . Black girls are much more likely to be incarcerated than white girls . . . and recent research does suggest that Black girls in juvenile detention are especially vulnerable to depression.[4]

This is a crisis! We are not taking care of our children. And if there is any one good reason to confront the racism and sexism Black women and girls uniquely endure in our country, it is this: "there is a correlation between experiences of racial discrimination and signs of depression among Black teens."[5]

The implications of this fact are profound. We know, for example, that Black teenagers can face, on average, five racially discriminatory experiences every single day. We know that those discriminatory experiences erode their mental health and put their lives in jeopardy. As the mother to two teenage girls, I can personally attest to the dramatic impact bullying and discrimination have on our children's mental health, well-being, and happiness. Many of our children are living with a perpetual stress response that makes them less well, inside and out, while jeopardizing their very lives.[6]

If we are to see the incredible progress of Black women continue, then we need to pay more attention to the state of Black girls. They are our future and a national treasure. They are our nation's future supersheroes. It's time we wake up.

I recently had a conversation with Adjoa Asamoah, a social entrepreneur who is leading a movement to tackle head-on the adultification of our children and discrimination against our people. Asamoah leads the CROWN Coalition, which is organizing a mass movement to pass the CROWN Act. "CROWN" stands for "Creating a Respectful and Open World for Natural Hair," and the act aims to protect Black people of all ages and genders from the anti-Black discrimination they face at school, at work, and in the community for how they wear their hair. The act recognizes a hurtful, essential truth: in today's society, Black women's hair is 3.5 times more likely to be perceived as unprofessional and Black women are 83 percent more likely to be judged more harshly because of their looks.

Asamoah has experienced this reality firsthand. She has faced discrimination when applying for jobs and has heard countless stories from Black men and women who have either been denied employment or terminated outright for their hair. She has heard horrifying stories of Black girls and boys disciplined—even arrested—at school for their hair. She speaks passionately about Deanna and Mya Cook, twin girls, age fifteen, who saw escalating punishment at school for their refusal to remove their hair extensions, which apparently represented a "uniform infraction." Or Andrew Johnson, a teenage wrestler who was given an ultimatum by a white referee to cut his dreadlocks right before his match or forfeit the competition.

Since she was young, Asamoah has for years organized at the state and federal levels to build momentum for the CROWN Act. California State Senator Holly Mitchell first introduced the bill in California in 2019, which was signed into law by Governor Gavin Newsom. Senator Cory Booker and Rep. Cedric Richmond are championing the legislation at the federal level. The CROWN Act is law in fourteen states and thirty-two municipalities.

The Act is so important because it empowers us, as Black women, to show up authentically. In my twenty-four years as a White House reporter, I have never been able to show up authentically—rocking my braids, for instance—because to do so would incite backlash and elicit bias, which would harm my career. If I were to wear braids in the White House, for instance, I would confront insinuations that, somehow, I am "militant" or "performative." This is ridiculous. Hair discrimination is race discrimination and it is a long-standing, legal practice that further codifies racism. As Asamoah reminds us, discrimination in any form is ultimately a net loss for everyone.

"Racism is a pretty bad business model. The research shows us that creating work environments that are truly inclusive, where people can show up in their own fullness and be who they are, are work environments that are more productive . . . (The CROWN Act) protects people to show up as who they truly are."

Imagine the productivity and engagement lost because of bias and discrimination. Asamoah is leading a movement to unleash the potential of Black people everywhere by creating an environment in which we can finally be ourselves.

TOMORROW'S GAME CHANGERS

As the Bible reminds us, we should never forget that our best change agents reside inside our home and in our communities—they are the next generation. The composition of the next generation is increasingly "nontraditional," which is code for stepping out from the shadow of convention. This next generation is increasingly representative of our community and our society's diversity. The next generation of formidable Black women aren't found only in the traditional, predominantly white institutions like the Ivy League. They are emerging from everywhere.

This is a welcome change. Years ago, when I was starting my career, Ivy League colleges and universities, like Harvard or Yale, would tell their students—and the world—that those special students were the ones to change the status quo because of their pedigree. For generations, if you walked the hallowed halls of prestigious, elite institutions, then you were anointed the promise. The hope. The change. Some of our role models followed this path. Think Barack and Michelle Obama, two Harvard Law graduates with Ivy League undergraduate degrees to boot (Columbia and Princeton, respectively).

This isn't to disrespect the former president and first lady. Not at all. But it's worth highlighting that so many incredible Black women have not emerged from these more traditional structures. Increasingly, the promise, the hope, and the change are emerging from institutions and communities that have been historically overlooked. Unsurprisingly, I see more of us emerging from our historically Black colleges and universities, long punching bags in public opinion and state and local budgets.

While the first historically Black colleges and universities (HBCUs) emerged in the late 1830s (the first being the African Institute, now Cheyney University of Pennsylvania), the majority of the HBCUs were founded in 1867—two years after the liberation of enslaved people and the conclusion of the Civil War. These were the deeply important and poorly understood Reconstruction years when newly freed Black men and women laid the foundation for the modern civil rights movement.

With slavery abolished, Black men and women throughout America realized that the fight was now for civil rights—namely, The Vote and basic investments in education. Many HBCUs were founded in the states of the former Confederacy, where the schools were granted land from the federal government and encountered entrenched opposition almost immediately at the state and local levels. Their mission was desperately important: throughout American history, Black people were not allowed to attend most of the predominantly white institutions of learning; where they were allowed, their attendance was subject to quotas, limits, and disparate treatment.

The early advocates of HBCUs and the countless number of students who bravely enrolled and pursued their education continued the fight for education for Black people everywhere. From their places in the world of higher education, HBCUs now lead the charge to change the country. Vice President Kamala Harris is a proud graduate of Howard University. Dr. King hailed from Morehouse College. Oprah Winfrey is a proud graduate of Tennessee State University. Superstar lawyer Thurgood Marshall graduated from Howard University law school before starting the NAACP Legal Defense and Educational Fund and, eventually, sitting on the US Supreme Court.

The list goes on.

From this cherished heritage, the next generation is already making waves. As I write this, a group of Black students—many of them women—are occupying Blackburn University Center at Howard University, holding the administration to account for poor housing conditions on and off campus. Elevating the most powerful practices from the nonviolence movement, the students are sleeping in tents to call attention to the mold that lines the walls of their dorms, the absence of COVID-19 testing, and general safety concerns. Indeed, as Frederick Douglass reminded us long ago, "Power concedes nothing without a demand." As the Rev. Jesse Jackson said to the students camped out at Blackburn, "Never surrender."

Some of us emerge from the humblest of circumstances—returning home from prison or entering the workforce after skipping college altogether. Life doesn't always give us equal chances and opportunities. Yet no matter from where they emerge, Black women still find a way to contribute, to lead, to make a difference—for ourselves, for our families, and for the world. Remember, one of our superpowers is transforming hardship into healing and pain into progress. You see this most clearly and most painfully in the criminal justice space.

Gina Clayton-Johnson founded the Essie Justice Group to advocate for Black women who have loved ones incarcerated, something Clayton-Johnson faced personally. She understood that the epidemic of incarceration that has disproportionately targeted Black men had another consequence: the suffering of the Black women who loved them and who bore the heavy responsibility of keeping families and communities together. Predictably, she quickly discovered that there wasn't support for, a policy framework for, and a culture of investment in these heroic women. She set out to transform her pain into progress by creating an organization that fosters community among Black women who have loved ones incarcerated. She mobilizes that community to advance changes in policy and how the system operates.

Garrett Bradley, another incredible Black woman, documented in very personal terms this very struggle in the Oscar-nominated documentary *Time*. Throughout the film, Bradley follows Sibil Fox as she fights for the release of her husband from prison where he is serving a sixty-year prison sentence. By incorporating home video footage and with strong cooperation from Fox and her family, Bradley creates a heartbreaking portrait of the burdens Black women carry silently while our men are trapped in a vicious epidemic of incarceration and in the criminal justice system. While the men are imprisoned, there are children who need to be raised, communities that need holding together, and fights to fight.

Clayton-Johnson and Bradley are remarkable young Black women who through organizing and art make us pay attention—pay attention to the struggles and unique contributions of Black women as they heal our community's hurt.

Even so, many of our heroic young Black women are practically children when they first assume adult burdens and responsibilities for creating a better world.

As I write this, young Black women are risking arrest, every week, by demonstrating in front of the White House to protest against the administration's inaction on voting rights. These young people followed in the footsteps of civil rights giants, leaders like Martin Luther King III, to hold their Democratic agenda to account for movement on voting rights.

These young Black children and the Black women guiding them on the front lines appreciate the vital importance of voting rights, and it shows in their attitudes and actions. As Cornell Belcher, Democratic pollster for Brilliant Corners Research & Strategies, said, securing voting rights is "the number one issue for Black women." The face of the fight for voting rights over the decades has been Black America; however, restrictive voting laws affect everyone of all races and genders.

Arndrea Waters King, the wife of Martin Luther King III, highlighted the important role Black women play in maintaining America's democracy. "Black women will continue to show up. We will continue to show our power," Waters King told TheGrio. "But now it's time for America to show up for us."

Tina Turner sang "We Don't Need Another Hero" in the movie

Mad Max Beyond Thunderdome. The most memorable lyrics emphasized, yes, "we don't need another hero/We don't need to know the way home." The hero, you see, is already in our own home and family. And she's young. And she's here, right now, waiting for you to pay attention.

UNSUNG HEROES, FOOT SOLDIERS FOR THE MOVEMENT

Black women, young and old, have long served as the unsung heroes of the civil rights movement. Black women have always set about the business of changing the world matter-of-factly. If there's work that needs doing, they do it. If foot soldiers are needed on the front lines, they do it. If there's food to cook to feed all the tired, weary activists in between actions and demonstrations, they do it. If the movement just needs bodies—to absorb attacks from police guards and water that cuts from fire hoses—it's often young Black women who answer the call.

Why do they do it?

For starters, we listen. Young Ella Baker, the legendary organizer, recalled listening to her grandmother's stories of and personal experiences with slavery. We respect our elders and whether it's at home or in the church, we listen. We pay attention. We absorb the lessons of our ancestors. At an early age, we develop an urgency to act.

Second, we have to. Black women are smart. We are observant. We recognize the cycles of ego and gridlock that can define our politics and activism. We appreciate that truly transformative change follows sustained, selfless contribution. We understand that without selfless leadership and a willingness to do whatever it takes, a lot of progress won't just happen. We can't afford for it not to happen. We adjust. We sacrifice. We serve.

Finally, Black women are unique in our mind-set. From an early age, we have a "collective orientation" that defines our leadership style. We live by the old adage "all for one, and one for all." We have the most marginalized identity in a society that often refuses to recognize our humanity, let alone offer us a seat at the table. Therefore, we understand more than most other people that the best way to fight oppression is to do it together and to win for everyone. Entrenched power doesn't concede easily and we win only when we are organized, coordinated, and advancing the interest of the collective. That mind-set has defined the leadership style of Black women for centuries, and when I study the next generation, it's clear that the orientation to fight for everyone is as strong as ever.

Our collective orientation drives us to pay attention to the gaps as we answer the question: Where is the unmet need?

The unfortunate consequence of this orientation is that so many Black women have been marginalized or written out altogether in our history. The truth is this: from the labor movement to the abolition of slavery to civil rights to the modern fight for Black lives, Black women have almost always played the role of operator and orchestrator of social change—selflessly executing the logistics in order to make our movements "tick."

Take Ella Baker, the legendary organizer who made possible so much of the civil rights movement's progress. She codesigned and implemented many of the plans set in motion by the Great Men of the movement while receiving just a fraction of the credit. When she came of age, she became active in the NAACP, first as a volunteer. She excelled at recruiting new members, mostly because she didn't talk down to the ordinary man or woman struggling to get by and provide for their families.

She also didn't put her leaders—who were almost entirely men—on a pedestal. She didn't glamorize their fancy educations or impressive credentials. She understood as all Black women do that our strength is in the collective—the "rank and file" who participate in the marches and demonstrations at great personal risk with no promise of glory or recognition.

Baker held her brothers in the struggle accountable.

Baker found Walter White, the former head of the NAACP and a prominent figure in exposing the horrors of lynching in the United States, to be highly egotistical and undemocratic in his leadership style. His behavior motivated her to fight to change how the organization operated.

Throughout her decades with the NAACP and particularly during the Walter White era, young Baker pushed the NAACP to make decisions more democratically and to involve local chapters and communities in the organization's decision making. She fundamentally changed the character and culture of the NAACP and did so quietly while its leaders, all men, reaped the benefits of cultivating the followers.

Baker understood that too many leaders, especially men of all races, project their self-image ostensibly to advance the mission but they do it to elevate themselves, at least in part. She appreciated that Black women necessarily work in the opposite way, that is, for the community, because a Black woman's image isn't welcomed—we're expected to stay in the background. From the shadows, Baker quietly knew how to drive change both inside and outside of institutions. She did what all young Black women do by not conflating love and loyalty with uncritical or blind support—like the young Black women at Howard who honor the brave and courageous legacy of HBCUs by also insisting that they can and should be better in serving all their students.

We can't underestimate the strength of our children. Sometimes we neglect to understand how they view the world and just how differently they see the world, often for the better.

Think for a moment about the horrors and the dangers of the COVID-19 pandemic. It was in that context that our kids risked their lives and defied death to march for what was right.

As many older Americans watched from the sidelines in horror as more Black men and women were summarily executed by the police, the young people left the safety of their homes to march in protest against the unjust killings of the innocent. They marched against the vigilante murder of Ahmaud Arbrey, against the knee on the neck of George Floyd, and against the no-knock warrant that led to Breonna Taylor's senseless killing. They marched for all of them. They marched in memory of Emmett Till and followed in the footsteps of generations of young people before them who made their voices heard at precisely the most inopportune moment because it was the right moment.

They did this in scores. Truly remarkable.

I didn't march. As a journalist, participating in that kind of visible activism would invite attacks of bias and present a conflict of interest. Nevertheless, my own children couldn't care less about my job. They simply craved justice for people who could have been their father, their uncle, or grandfather. Or themselves.

Against any qualms about COVID-19 or my career, they marched—and they marched—and they marched. They were determined. I could not stop them.

My daughters marched with hundreds of others from the Baltimore area—children of all races, all genders, and all walks of life. My fear for them quickly gave way to pride as I saw my babies stand up for justice. My daughters—one of them a senior in high school and the other a seventh grader—standing up for something greater than themselves. With their masks on, they were nameless. They stood up and led, just like those who have come before. History has shown us, over and over, that it takes a normal person with courage to truly make a difference. It's not the always the visible heroes. It's the unnamed, many in our own homes, who come out en masse to truly make a difference.

My daughters follow in a line of young Black people, many of them women, who have changed the world.

I think about the Little Rock Nine:

Minnijean Brown, Elizabeth Eckford, Ernest Green, Thelma Mothershed, Melba Pattillo, Gloria Ray, Terrence Roberts, Jefferson Thomas, and Carlotta Walls.

Recruited by the local president of the Arkansas NAACP, these nine students possessed extraordinary strength, courage, and conviction to sign up for the tough, nearly impossible task of integration. Leading up to the start of the school year, the young people underwent intensive counseling and therapy as part of their preparation.

They needed it.

After all, these students stared down a literal army. Just two days before the start of school, then governor Orval Faubus ordered the National Guard to ban any Black student from entering the state's schools, claiming it was "for their own protection." When the students integrated the school on September 4, 1957, they were met by furious white students and adults who protested their presence with racial epithets and spit. They stood face to face with National Guard members who, under direction from their governor, refused to let them enter.

The mob protesting against their presence would only grow. Days later, it would swell to more than a thousand when the students tried to enter the school once again, unsuccessfully. It would take the deployment of federal troops by the president of the United States to see the Little Rock Nine successfully escorted into the high school.

I think about little Ruby Bridges, who integrated an elementary school in New Orleans. Ruby was born in 1954, the year the US Supreme Court decision *Brown v. Board of Education of Topeka* came down, declaring school racial segregation unconstitutional. She was just six years old when federal agents were sent to New Orleans to accompany her to school. She was the only Black student in the entire school. And every single teacher, except one, refused to instruct her. She would eat lunch alone. Participate in recess alone. Take her studies alone.

Her life was threatened time and time again. She suffered through intimidation tactics (one white woman, for instance, put a Black baby doll in a coffin and left it outside the school). She endured nightmares. For large stretches of time, she lost her appetite and refused to eat. The stress was barely tolerable.

Six years old.

"A little child shall lead them."

Ruby Bridges persisted. She integrated schools in New Orleans all the way through high school.

This isn't ancient history. As of this writing, this was just over sixty years ago. Many Americans alive today were alive then.

And it was the children who had the courage to do the hard

thing—the uncomfortable thing—to make a difference. It's important to maintain that perspective because our young people are acting for our national interest every day. We often overlook them, equating their youth with inexperience. Assuming their efforts are simply symbolic gestures. But that does them and us a disservice. For the entirety of this country's history, it has and always will be the young people—the nameless, faceless young person—who are the vital change agents.

It was the young people who spearheaded the movement for abolition in the eighteenth and nineteenth centuries. Take Sarah Parker Remond, a young Black woman who helped ignite the movement for abolition. She gave her first speech at sixteen years of age. By the 1850s, Remond was delivering impactful and important speeches in Europe, alongside orators such as Frederick Douglass, to mobilize international support for abolition.

It was the young people who started the labor movement, also in the nineteenth century. Girls as young as ten years old organized and protested against the deplorable conditions women were subjected to in the textile mills throughout New England. When one of their leaders was fired, eight hundred young women walked out, striking in protest. Their actions represent one of the earliest examples of the organized labor movement.

More recently, when Trayvon Martin was murdered—a tragic, heartbreaking moment widely recognized as the start of #BlackLivesMatter—it was the young people, specifically, thousands of elementary, middle, and high school students in Miami, where Martin had gone to high school, who walked out of their classrooms when no charges were first filed against George Zimmerman (the man who shot and killed Martin). The "Dream Defenders" walked for three days to the police station in Sanford, Florida, to demand an audience with the officers who refused to charge Zimmerman. Their actions changed the course of that case and altered history. The origins of #BlackLivesMatter were seeded in those actions. Those young people changed the world.

They always have.

RISING TO THE OCCASION

"Sometimes we choose our problems and sometimes our problems choose us," said Tiffany Loftin, the national director of the NAACP Youth & College Division. She would know.

Loftin was raised in comfort and relative affluence in Los Angeles. Her parents were union members in the entertainment industry. During her early years, life was good. That all changed in an instant when her parents divorced because of domestic violence. Suddenly Loftin went from living in a very large house in north Hollywood to living in a shelter in Sherman Oaks, California, with her mother. Newly enrolled in public school, Loftin looked after her two younger siblings while she worked hard to earn admission to college.

A first-generation college student, Loftin first started organizing at UC Santa Cruz when the California Board of Regents raised the tuition, making it more challenging for other first-generation students like Loftin to stay in school. Loftin fought back and she hasn't stopped since. In the struggle as she transforms pain into progress, Loftin discovered her "why":

> My "Why" in this moment is because I recognized very easily in my college experience that there are so many things that we deserve. So many things that are rights to us. The right to education. The right

to vote. The right to safety. The right to be able to choose what I'm going to do with my own body. The right to a great, incredible union job. The right to travel freely. All of the things that I think about. The right to have your votes reinstated after you're incarcerated or go through the judicial system. And there are so many people who make up systems who try to stop that and I found my voice in college. I found the power of community organizing in college. And I found family in the people in that community. And I haven't stopped since I think it is my God. My therapist calls that my destiny package. I think it is my God-given destiny package on this planet to do this work.

Like Ella Baker before her, Loftin connected with her ancestors—seeking inspiration and an unwavering community through them.

I think that our strength and magic are rooted in that process, in that journey to finish what our ancestors began and to save everybody else. I know that's not fair. I know that we don't want to do it. I know it would be easier to do anything else. But if our superpower is that, and nobody else can do it, then I am fine accepting that responsibility from God.

Like the civil rights greats before her, Loftin understands the power of love as the fuel for change. Love is what propels us to persist. Love, to paraphrase Dr. King, is what drives the darkness out. It's the only thing that can. When we are too tired to carry on, love brings us back center and motivates us to keep pushing. As Loftin says:

Us fighting in spite of, as us loving in spite of. Not giving up in the process. I have lost friends. I have lost friends who have taken their lives in this work. I have lost friends who have been killed by the hands of police and are still incarcerated because police. The war that we are fighting is bigger than us. And there is a spiritual command to do this work. And the amount of love that we have in spite of it is really our saving grace. We wouldn't be able to do it if we didn't have that, and if we didn't have that, we'd be them. And I'm fighting them. So, I can't be like them.

The next generation, like our ancestors before us, understands the importance of fighting on. Through our ancestors, we find healing too.

CHAPTER 8

Healing

> *Healing is an act of communion.*
> bell hooks

THE JOURNEY HOME

As the elders shout out in praise in the Black church, a classic Pentecostal experience, they call out the names of the elders and ancestors who have laid the groundwork for change in preparation for the next generation of standard-bearers.

If my name is ever mentioned in that roll call, I want the next generation to know that the fight is bigger than any of us. In the end, it is about helping others. When I reflect on my own life and reflect on the question "Have I done enough?" I can't help but think of occasions when my questions and my pursuit of truth have moved the needle on an important issue.

One moment still brings me to tears. I asked then President Trump if he would apologize to the Exonerated Five for calling for their execution, in light of Ava DuVernay's recent movie *When They See Us*, which told the full truth of their coerced confession in the beating of a jogger in Central Park.

Trump merely said that the kids had confessed to the crime and that there was no reason to apologize.

Years later, I met The Five, who were thankful I raised their names to help clear the misdirected efforts by Trump, and others, to have them executed. Even though the news cycle had moved on at the time of my question, justice clearly still mattered. It mattered that the truth was told. It mattered that the powerful were held to account. It mattered that some modicum of justice was eventually served when their convictions were vacated and the City of New York forked over $41 million to compensate for the vicious, discriminatory prosecution decades earlier.

The case of The Five raises an important question of healing. Black people, including and especially Black women, carry so much pain, too much trauma. One area where I am most excited about the next generation leading is the cause of our healing.

Sure, Black Women Will Save the World—but will we save ourselves?

I had this conversation with Tiffany Loftin, and we bonded over our shared journey to Africa in our individual pursuit of healing. Loftin tried to find her solace in Ghana, where she encountered the slave dungeons that had imprisoned our people centuries ago. She had this to share:

> I have been fighting and fighting and fighting. I have been doing this work for twelve years. I started being an activist and an organizer in 2007. When I went to Ghana, you are taken to, as they are called, the Slave Castles, the Slave Dungeons. I had been anticipating this moment, the entire trip, right? It is my second time in

Africa, but my first time visiting Ghana, and I'm anticipating this moment of being at this castle and just feeling the spirit of God; my ancestors were going to hug me and love on me. I was going to somehow get more answers. It is almost like when folks go to Palestine, and they get to meet at the places where Jesus wept and where he died. And I'm like, "This is my Damascus moment."

As Loftin replays her epiphany, she is so excited. Like so many of us, she comes from a family that doesn't know their genealogy beyond the grandparents. Loftin had exhausted all possibilities on platforms like Ancestry.com or the National Archives. Going to the site of the Slave Dungeon was her only chance at clarity—and closure. She continues:

I am this young Black girl, I'm on this trip, and I'm superhyped. I get there, and our tour guide's name was Justice. So powerful, Justice walks us through this castle, this dungeon. I get into this dungeon, and he says, "The men and the women were separated." I walk through the dungeon that is dark, cold, moist because the building is on the coast; it was entirely wet (just a) chilling dungeon.

I stand there almost as if I expected an alien to come and warp me up; I was expecting this huge force to speak to me in a place and moment where I thought I would have gotten more answers. Questions started to come up. I said to myself, "Okay, did my ancestors come through this one?"

Loftin's questions quickly led to more questions.

Well, I don't know if they were man or woman, or if they were separated. Then they tell you that there are about sixty castles on the coast. I'm like, "Well damn. I might be at the wrong one. Were they younger? Were they older? Were they queer? Were they pregnant? Was it in Ghana, or was it a different country?"

Loftin's questions expressed her desire for connection and, ultimately, healing. But the experience was disorienting. It left her, at least initially, with not just more questions but pain as the trauma of her ancestors' experience became so very real in that moment.

> I left with more questions and a little bit more emptiness. I started crying. This guy named Michael—I'll never forget their names—this man named Michael came over, and he hugged me, and he rubbed my shoulder. He was part of the tour staff. He said to me, "Sister, I know this hurts. I know. But I hope through the pain, fear, and confusion that you're feeling right now, I hope that you feel a sense of accomplishment." I'm still crying, and I can barely hear him because I cry ugly. He says, "You were able to complete the journey that your ancestors died for. You've brought them back home. So even though you don't feel like you got the answers, you have brought your ancestors back home in this journey of life that you didn't even know you were being used and operated on this journey to do that. But you're the first in your family to be able to do it."

You were able to complete the journey that your ancestors died for. You brought them back home.

There is so much power in Loftin's story. That power is both particular to her and universal in the African American experience. I've also traveled to Africa to visit the Slave House, Maison des Esclaves in Senegal on Gorée Island. I've been there several times.

I did the exact same thing as Loftin.

I cried. And I cried. And I cried.

On one trip, we participated in a liberation ceremony. Like Loftin, I was searching for a sense of "I'm home." I desperately wanted to feel a sense of connection to the place that played such a horrifically powerful role in the history of my people and our people. But like Loftin, I left the experience feeling empty. I didn't walk away thinking "this is it." Instead, I just I didn't get it.

After one trip, I did a story and discovered that many African Americans made a similar journey to find out who they were. They too walked away with more questions than answers. Many became depressed because they didn't find the answer. Some were overwhelmed with despondency. A few, very tragically, committed suicide.

I empathize with their pain. Like Loftin, I have spent hours scouring sites like Ancestry.com trying to find answers—trying to find healing. With every insight uncovered, you just want more. I found out, for instance, that my great-great-grandfather was sold on an auction block at a fair in North Carolina. I dug, and dug, and dug to find more—but there are no more records, at least none that I can find. I think of him often, I consider the terror he must have felt in that moment. I reflect on his courage to persist, not for himself but for future generations—for me. I want more. But I have come to accept that some things are simply not knowable and, ultimately, we must heal ourselves.

HEALING OURSELVES

We have spent so much of this book discussing the ways in which Black women are confronting the world's biggest problems. We've explored the ways that Black women are uniquely challenged as they do so. But it's fitting that we spend time elevating the ways that Black women are healing themselves and disrupting the narratives that insist that our collective future cannot change for the better.

All over the country, remarkable Black women are discovering ways to heal our community and write new narratives about what's possible for Black women: new hope, real healing, and good health.

This conversation is overdue.

I recently met two remarkable social entrepreneurs who are writing the next chapter for Black women across the world: T. Morgan Dixon and Vanessa Garrison, the cofounders of the health nonprofit GirlTrek. They get it, as Garrison shared:

> We hear from—and I know because I am one—Black women that we are living in a pressure cooker. A place where from every direction it feels like there is relentless attack against our survival, against our livelihood, against our joy, and Black women are at the center of the keeping together of the families through the midst of that in the pressure cooker. We're in the midst of the keeping together of community through the midst of that. We're in the midst of the keeping together of organizations and movements in the midst of that. And for me, GirlTrek's work is to understand the environment by which we live to provide a pressure point, a relief to some of those pressure points through community, through practical solutions, through courageous conversations and to acknowledge for Black women that even though what we are experiencing now feels traumatic and is traumatic, it's not unprecedented, and we do have a documented history of how to overcome and how to move through and how to get to the other side.

The remarkable women of GirlTrek have catalyzed a movement of more than one million Black women who walk every day to create space for their own healing and to send a message that the health and well-being of Black women deserve a statement of solidarity. We don't march only for the liberation of all of us. We also march for ourselves: our health, our happiness, and our sense of peace. As Dixon shared in our interview:

> Violence threatens to choke us and our voices back every single day. And that every time I open my front door, it is a declaration and a discipline of hope. And what I'm saying is [for] a people who have experienced population level trauma to hope is radical. And so, this kind of physicalizing hope is what GirlTrek is. It's not about weight loss. It's not about fitness. We're not walking enthusiasts. We are asking people to walk on faith every day that it is worth living, that this life is worth living. So that's it.

When I asked them why they started GirlTrek, Garrison responded with her powerful origin story:

> I couldn't locate myself in any of the healing spaces that were available to me. I couldn't locate a space that was speaking to me as a young Black girl whose mother had been in prison, whose family was making a dollar out of fifteen cents, who was trying to maintain my level of blackness would it be, but also progress into a space of the unknown. I was, when I met Morgan, just like nineteen years old and flailing. And yet I knew that something needed to change so that I could change the trajectory of my life. So, for me, it's always been a very personal call around how do I transform my

own traumas? How do I do it in a public way so that other people can learn from them?

How do I give other Black women permission so that they can do that same thing? Because I think it's a winning formula. I think that when we pull the veil off what we are struggling with, when we have those conversations boldly and publicly in the way that GirlTrek has for the last ten years, that's what works for us. That's where it started; it started with Dixon and me having bold, public conversations about what we were going through in our own lives, and we did it through weekly emails and we did it through our social media accounts. We were just authentically talking about things that Black women were coming to us and saying, "No one's talking about that." Still to this day, with all the media outlets dedicated to Black women, with all the great creatives, with all the social media connections, there's still this space where a lot of Black women feel left out, even if they can't put their finger on why that is.

They feel not seen, they feel not supported, they feel as if they are on the outside, and that's the thing that fires me up. How do we make space so that those women are, like, this feels like home? This feels like home for me. I think that when we can make women feel that way, when you can make a woman feel that level of home, she then feels safe. When she feels safe, she feels less fear. When she feels less fear, she feels like she can make bold and courageous choices in her life. That's what's happening with GirlTrek.

Ultimately, the women at GirlTrek walk to breathe—to reset—to gather themselves because the daily existence of Black womanhood is hard. They have built a movement to uplift in the culture in which Black women need intentional space for our own health and happiness to not only survive but thrive. I related the GirlTrek ritual of walking to my annual sojourn to Martha's Vineyard. While that trip may sound very bougie to some, I am unapologetic: I need it for my peace; I need to reset. I have rituals for when I am there that serve that purpose. For instance, every year I participate in Polar Bear. I get in the icy water enveloped by that circle of blackness and in my that moment, I connect to other beautiful Black people and our shared intent: healing. It's a spiritual experience.

Wonderfully, I am not alone when I am there. Like the strong women at GirlTrek, I find community and communion in my healing. This last summer, I found something special with Keisha Lance Bottoms as we visited on her porch. We talked about everything. And we talked about nothing, including the skunks nearby. The substance of the conversation mattered far less than the very fact that we had the time and space to have it. To both just be and to be together.

When I shared this with the ladies at GirlTrek, they resonated with my experience as I had with theirs, with Dixon sharing:

> GirlTrek does exactly what you just described . . . which I'm just like, okay, invite me to Keisha Bottoms's porch, please. But sitting on Keisha Bottoms's porch where you're actually, you said it, four times, it's peaceful, it's peaceful, it's peaceful, it's peaceful. That we as Black women do not have enough opportunities to build community that is separate from our traumas, our work, our everyday hustles in this way, that GirlTrek is like, we are going to do this unapologetically, and women essentially experience that sitting on the porch with each other through walks, right? They experience those conversations, that intimacy, that level of peace. We've almost

scaled that peacefulness of Martha's Vineyard that you're talking about and taking it to women so that they can access it too in their communities, through this discipline. It's really a spiritual discipline and to come together with other Black women and have that.

What I learned from the women of GirlTrek is that time together is important because we, as Black women, need to bear witness to what is happening to us, individually and collectively. They reminded me of a truth first memorialized by bell hooks: healing is an act of communion. The Rugged Individualism of America is mostly myth and it's particularly out of step with the needs and realities of Black women. We are collective in our leadership, we are collective in our contribution, and we are collective in what we most need: each other.

They also reminded me that there is spaciousness in our community because it affords a degree of anonymity. Alone, Black women are always marked—we experience bias and discrimination; we are fetishized; whatever the particular experience, we are marked. But when we are together, there's an element of freedom in that. Dixon shared the importance of spaces created for, by, and with Black women because it's likely the only truly safe space many of us ever encounter:

Each individual Black woman I know is phenomenal and she's carrying the weight of her phenomenal-ness every single day. And sometimes we just need to sit on the porch where we're just with our friends. Where we're just in safe space. And so, for me, when I put on the GirlTrek shirt, I went into a community center, one time to go to a GirlTrek event, and these little college students was like, "Hey, welcome sister. I want to introduce you to GirlTrek. It's a national movement." And they gave me the clipboard and I just signed up April. Cause I didn't even need to say nothing. I just signed the thing, and I was like, y'all do a good work and I just kept moving.

And I'm glad. There's something about putting on an identity of hope that doesn't require you to struggle to put your worthiness into the world. And for me, putting on a GirlTrek blue shirt is putting on an identity of hope that is a collective hope. That says, 'we've come too far to give up now' kind of hope. That says, 'this road ahead ain't nothing compared to the road we've traveled' kind of hope. And it does. I love that.

As they look ahead, the women of GirlTrek are expanding their work internationally and to effect systems change. By promoting a culture of good health and healing, they hope to extinguish the culture that takes Black women for granted—they ultimately want to disrupt the very system that normalizes our shorter, less healthy lives. This is a new narrative focused on us. And it feels good to think about what's possible if we seize their vision, together.

When I reflect on what the women of GirlTrek have accomplished, it's staggering. They aren't slowing down. They are building a movement dedicated to changing the very systems that have prosecuted Black women for centuries, as Dixon shared:

Our first goal was mission accomplished. We wanted to get a million women walking in the same direction every day to save their own lives. The next goal is more concrete in systems. The next goal is, how do we increase the life expectancy of Black women by ten years in ten years. We're dying a decade earlier than every other woman on the planet and so that's a scary one. That requires us to work with mayors. It requires us to work with city engineers. It requires us to work with a coalition of people who know how to do this and that we hold them to the fire to change the system so we can live. That is the pathway ahead. And we'll start next year with going back to grassroots organizer and we're going to have a groundswell of organizers through the streets with behavior change, and then we are really going to be starting to work on major coalitions. What Vanessa's talking about really does matter. We have now I think about the power behind us, these women behind us who have audience with people who really are powerful (and capable of effecting) systems change.

This is good and worthy work because the women of GirlTrek are building a movement dedicated to our collective and individual healing and dismantling the very systems that put us here. This work is long overdue and represents a culmination of sorts for Black women leaders and activists everywhere. After decades of saving the world, Black women are building the infrastructure and the movements dedicated to our salvation and well-being. We are finally naming our health and well-being as a cultural priority and leveraging our superpowers to deliver for ourselves and our children.

It's time.

CHAPTER 9

What's Next

WE MATTER

We have heard from so many of our leaders throughout this book. From the grassroots to the "grass tops," Black women are changing the world. Like never before, we are powerful: Black women lead our government. We run our institutions. We marshal movements. We have taken our seat on the Supreme Court. Never in our history have Black women played such a vital and dominant role in the course of national events as they do today.

With that great responsibility comes more of everything: scrutiny, pressure, expectation.

The only thing that has not followed is a deep and thoughtful study of what it's like to be Black, female, and powerful. While countless stories, biographies, films, and television series have been devoted to men of all races and white women, Black women very rarely find themselves at the center of such careful study—let alone portrayals in which we are celebrated or even exalted.

If this book achieves anything, I hope it normalizes this: pay attention to Black women and girls.

As we become more powerful and represented at all levels of our society's decision making, we should all care how Black women navigate a society that wasn't built for us. Even as we try to change America for the better, Black women grapple with a particular set of challenges that are unique to who we are: double standards at work, a culture that refuses to see us without hostility, structures and institutions that actively work against us, from health care to education to the worlds of politics and power. As we have seen, those challenges begin early. Black girls endure a truly horrifying degree of "adultification" that leads them to more punishment and less care and investment.

Understanding the challenges confronting Black women and girls is important. It should compel at least of some of us to try to address the structural and cultural forces attacking Black women once and for all. Doing so will make us a better country, which will benefit everyone. A Black woman unleashed is a change agent like no other.

Paying closer attention to Black women also reveals an obvious and important point: we are human! Despite our heroics, we are not perfect.

Black women, like all people, are complex.

We are flawed. We are imperfect. We are doing our best as we negotiate the change we want to effect in the world as it is. Our lives are often a difficult, delicate dance. Our failings, where they exist, should invite grace and a recognition of our humanity. After all, we are all part of one human family—we just want to take our rightful seat at the proverbial table and have our humanity acknowledged, alongside yours.

I hope that came through too.

As I close this book, I thought it fitting to elevate the voices of our young people. In the end, their thoughts, opinions, and perspectives matter most. They are, after all, What's Next.

WHAT'S NEXT

What's next? Well, The Change is what's next. That's where you, my dear reader, come in.

To start, this book outlines some, but not all, of the obstacles and barriers Black women face in their lives. It also shows all the ways that Black women transcend those challenges to serve and protect America in one of her darkest hours. If Black women are going to continue to save the world, then it is time that we act together to uplift Black women and dismantle the many obstacles facing us.

Let's get to work. Here's where you can help:

- *Acknowledge Black women.* "Erasure" is real and dangerous. In particular, take the time to learn about and acknowledge all the ways that Black women have advanced social justice across society—from the abolition of slavery to civil rights to our modern political era. We have played a huge role in the progress of this country. Help us own our contribution by lending your voice to our acknowledgment. Do this in big and small ways. When a Black woman is driving a successful effort at work, acknowledge her. Equally important, do your part to elevate the struggle and pain of Black women, and trans and nonbinary individuals who go unseen in our society. Remember: one hundred thousand Black women go missing every single year. Many are victims of violent and sexual crimes. Trans and nonbinary people are victimized, even murdered, just for existing. Nevertheless, our suffering is very, very rarely covered in the press. Help us change this. Our pain matters. Together we can do something about it.

- *Stop discounting Black pain and suffering.* It's killing us. Literally. From our health care system to mental and physical health in the workplace to our lives at home, listen to us as well as respect our feelings and experiences. Stop either overlooking our very existence or perceiving us as superhuman. It's not that we have a higher threshold for pain and suffering; it's simply that our experience is severely discounted. Do not conflate our coping with pain and suffering with a preference for them. Hear us when we say "stop" or "what you're doing is hurting me." Respect our bodies, and our personhood.

- *Own your bias and work on it.* None of us are perfect. All of us struggle with some form of bias. It's human nature, but it doesn't make it right. At work especially, Black women are too often viewed as "atypical," especially when we're contributing to leadership or in ways that are highly visible to others. In turn, the bias of our colleagues leads us to be judged and evaluated more harshly than others who do not share our race or gender. This all adds up to hold us back from that next opportunity or promotion. Check your bias and work on it. We deserve better.

- *Support Sisterhood.* Respect spaces for sisterhood and invest in it. There are tremendous organizations (for example, nonprofit organizations, recreational and student sports teams), networks and professional associations, and affinity groups that are devoted explicitly or implicitly to the notion of Black Sisterhood. Invest your dollars to demonstrate support. Organizations like GirlTrek are catalyzing entire movements dedicated to uplifting the health and well-being of Black women. This is important! After all, *sisterhood is our superpower,* and all of us should do what we can to nurture a culture of sisterhood in our communities.

- *Embrace new models of leadership.* Black women are leading in big and bold ways. As we climb in our professions, many of us are leading "outside of norm," refusing to conform to models of leadership pioneered by white men, in particular. That's okay! One is not better than the other. All we ask is that our models of leadership are elevated and respected, not discounted.

- *Respect our children and their childhoods.* "Adultification bias" is hurting our children. More Black girls are suffering in their mental health because of racism and sexism amplified and extended by social media. Pay attention to how you and your family might participate in the forces of racism and sexism. Seemingly innocuous gestures like asking to touch someone's hair are hurtful and contribute to a climate in which our children, especially our teenagers, feel isolated and targeted. Let our kids be kids—respect their personhood and personal boundaries.

- *Invest in our political strength.* Leaders like Stacey Abrams should not just be lauded after they've saved the day—they deserve investment. Let your dollars speak to your priorities. Support Black women in their political pursuits. Invariably, we are working *for all of us*—help us. There is at least one more glass ceiling to shatter.

- ***Let us speak for ourselves.*** Sometimes the best thing anyone can do to help Black women is to create space for us to speak for ourselves. Sometimes the answer is as simple as stepping back and creating space for a different point of view or inviting us to speak if others are taking up too much "air time." The small gestures matter as much as the big ones. Be an ally.

THE MOVEMENT

I owe a huge debt of gratitude to the significant number of Black women and girls who inspired, in ways big and small, this book. I've been fortunate to engage with some of our nation's most important leaders and a legion of emerging talent spanning the worlds of activism, politics, health care, and media. It should go without saying that the Black women highlighted in this book represent just a fraction of the Black women everywhere who are moving their communities forward. We should sit with the fact that such a comprehensive profile of Black women and their excellence is so unusual.

For your reference, I've included a list of every Black woman featured or named in this book as a way for you, the reader, to support their work or lift up their legacy. It is my fervent hope that #blackwomenwillsavetheworld will drive attention to and investment in all their worthy causes and memories.

They are listed below in alphabetical order, by first name:

Adjoa Asamoah
Adrienne Jones

Alexandra Hughes
Alicia Garza
Amani Allen
Ana Rodney
Angela Davis
April Ryan
Ayanna Pressley
Barbara Arnwine
Barbara Williams-Skinner
bell hooks
Christian Nunes
Cora Masters Barry
Eartha Kitt
Ella Baker
Erica Green
Fannie Barrier Williams
Fredrika Newton
Garrett Bradley
Harriet Tubman
Ida B. Wells
Isisara Bey
Josephine Baker
Joyce Beatty
Kamala Harris
Karen Bass
Keisha Lance Bottoms
Dr. Kizzmekia Corbett
LaTosha Brown

Lori Lightfoot
Maxine Waters
Maya Angelou
Meghan Markle
Melanie Campbell
Michelle Obama
Moya Bailey
Naomi Osaka
Nikole Hannah-Jones
Octavia Spencer
Oprah Winfrey
Ruby Bridges
Serena Williams
Sha'Carri Richardson
Shirley Chisholm
Simone Biles
Sojourner Truth
Stacey Abrams
Sunny Hostin
Susan Rice
T. Morgan Dixon
Tamika Mallory
Tarana Burke
Tiffany Loftin
Valerie Jarrett
Vanessa Garrison
Viola Davis
Yamiche Alcindor

These women have taught me that Sisterhood is our superpower. Together, we claim our seat at the table and change things for our people. I am indebted to them for their courage and continually inspired by their vision. Over the course of writing this book, I was staggered by the scale of leadership exemplified by Black women today. Our ancestors would be so proud.

A CLOSING NOTE ON VULNERABILITY

Black women are not allowed to emote or show that we are human. This is unique to us. We are not allowed to show weakness because that "little bit of blood in the water" invites a strike. We've learned the hard way, and I learned a long time ago: never let them see you sweat.

We are supposed to stand strong. We are supposed to bear it all. We are supposed to bear it for the children, for the community. Black women are not given the grace to show: "I need you."

We are the wet nurse. We are the sharecroppers in the field. We are doing it all. Once you shed a tear—that act of vulnerability—you risk it all.

When I worked for Boutique Radio Network (now my former employer), I could never show any outward expression of vulnerability. I was hustling hard then to establish myself and take my seat at the table as the only Black woman in the White House press corps.

Things got worse when management changed. In 2018, I had an important trip to New York scheduled. When I was on the phone with the new owners, I learned that my father was declining. I wanted to share what was happening. I wanted to cancel the trip. I wanted to ask for space, for grace, for a moment. But I couldn't. Then, the day before I was supposed to leave, my father died.

The company was in turmoil. Everyone's position was precarious. It was knives out. It was also clear that the new owners did not care about me or us as people—as human beings. As a result, I didn't have the grace or space to mourn. I had to keep pushing forward. In the most vulnerable moment in my life, I couldn't show pain or ask for help.

Later at the event, I shared what was happening with former New York Governor David Paterson who shared that he had done the same thing. Afterward, he simply said, "God bless you."

Those perceived as strong have to continue to be strong.

Our Black men are taught to be strong. Black women are taught to be twice as strong.

That's what we do as women, especially those of us who have been placed in positions to fight, speak, and question on behalf of the community. I don't take for granted the humanity of each Black woman who has moved the ball forward for our people. You may be thanked, but you aren't seen—they don't understand how much you take in to save the world.

I believe my womanhood and my elevation as a woman happened when my mother transitioned. I cried so hard, especially on the weekends when I was alone. I was pregnant with my daughter. I didn't show my pain publicly. But every moment of every day I thought about the words and guidance she had given me as I pressed forward. In our individual behaviors as Black woman, you can extrapolate the truth about our culture. We don't live in a culture that supports us to be vulnerable. For Black people, our vulnerability is mistaken for weakness. Why are Black women stereotyped as "hard"? Well, because we have to be. That's why.

One of the most vulnerable moments for Black women is when we take the mask off. When we remove our makeup, and that hardness that falls down in that moment we are vulnerable—we are real.

Never let them see you sweat.

After my father's death, Martha's Vineyard played a pivotal role in my healing. I made my father's funeral arrangements there in that place of solace, comfort, and community. In the end, it was a beautiful funeral. I am grateful that I had the space and the support to think about him, his legacy, what he meant to me, what he meant to all of us. In that special place, I had the space to be vulnerable.

...

In 2017, I received an honorary doctorate from Morgan State University, my alma mater. My mother worked there for forty-two years. I practically grew up on campus. It's my home. As I stood to receive my degree, a beautiful, warm breeze passed over me. I was overcome with emotion. That day—that moment—was perfect. But it was too much to process. Joe Biden was on the dais with me, standing nearby. He saw me and simply asked, "Do you need a hug?" I shook my head "yes."

He hugged me.

There are moments when people accept our tears. I hope this book ignites more of those moments. We are, in the end, one human family.

For my part, I am working on it. I am trying harder to be human, not just a superhero.

Acknowledgments

I begin by acknowledging all the Black women who lifted us when we could not lift ourselves, who moved us forward as others stood still.

I dedicate this book to my extraordinary family. I want to acknowledge my late mother and father, my daughters, Ryan and Grace, as well as my fiancé, James, who constantly shows love and exemplifies it as an action word. Thank you, all of you, for always having my back.

I am enormously indebted to the extraordinary Black women who participated in this book. Thank you for your example and for sharing your wisdom, insight, and pain to lift up lessons that only make us better. Our conversations have healed me, and I know they will help heal our readers. Thank you for helping the world to *see us*. I am so grateful for you.

Thank you to inimitable Senator Cory Booker for his foreword. Your unique seat and status shed a powerful light on Black women and our significance. Thank you for uplifting *this* conversation.

I must acknowledge the extraordinary Patrik Bass, my editor, who

ACKNOWLEDGMENTS

believed in this project when no one else did. Patrik, your vision and inspiration motivated this beautiful, soul-giving work. Thank you for being you.

I am beyond grateful for Traci Wilkes Smith and Peter McGuigan for their extraordinary representation and advocacy on my behalf. And I want to acknowledge and thank my friend Fagan Harris for his steadfast partnership throughout the life of this project. I also want to acknowledge Neyat Yohannes for her invaluable research and editorial support.

Finally, I must thank every woman who has poured into me, leading me to this moment. I am in awe of Melba Wilson of Melba's of Harlem. I stand on the shoulders of Karen Boykin-Towns, Hazel Dukes, and *every* woman who set the table for us to own the room and beyond.

Notes

Introduction

1. Inaugural Address by John F. Kennedy, President of the United States, January 20, 1961.
2. Langston Hughes. *Let America Be America Again.* New York: George Braziller. 2004.
3. Juanita J. Chinn, Iman K. Martin, and Nicole Redmond. "Health Equity among Black Women in the United States." *Journal of Women's Health* 30.2 (2021): 212–219. https://doi.org/10.1089/jwh.2020.8868.
4. Jameta Nicole Barlow. "Black Women, the Forgotten Survivors of Sexual Assault." American Psychological Association. February 2020. https://www.apa.org/pi/about/newsletter/2020/02/black-women-sexual-assault.
5. Janice A. Sabin. "How We Fail Black Patients in Pain." AAMC. January 6, 2020. https://www.aamc.org/news-insights/how-we-fail-black-patients-pain.

NOTES

6. Bianca Barratt. "The Microaggressions towards Black Women You Might Be Complicit In at Work." *Forbes.* June 19, 2020. https://www.forbes.com/sites/biancabarratt/2020/06/19/the-microaggressions-towards-black-women-you-might-be-complicit-in-at-work/?sh=38a42ae62bda.

Chapter I: The Superpower of Sisterhood

1. "Voting Laws Roundup." Brennan Center for Justice. Last modified May 28, 2021. https://www.brennancenter.org/our-work/research-reports/voting-laws-roundup-may-2021.
2. "Voting Laws Roundup." Brennan Center for Justice. Last modified May 28, 2021. https://www.brennancenter.org/our-work/research-reports/voting-laws-roundup-may-2021.
3. "Statement of Concern." New America. Last modified June 1, 2021. https://www.newamerica.org/political-reform/statements/statement-of-concern/.
4. Martin Luther King Jr. *Letter from the Birmingham Jail.* San Francisco: Harper San Francisco, 1994. Print.
5. bell hooks. "Sisterhood: Political Solidarity between Women" *Feminist Review* 23.1 (1986): 125–138.
6. Laura Morgan Roberts, Anthony J. Mayo, Robin J. Ely, and David A. Thomas. "Beating the Odds." *Harvard Business Review.* March–April 2018. https://hbr.org/2018/03/beating-the-odds.
7. Laura Morgan Roberts, Anthony J. Mayo, Robin J. Ely, and David A. Thomas. "Beating the Odds." *Harvard Business Review.* March–April 2018. https://hbr.org/2018/03/beating-the-odds.
8. Shannon Sales, Monica Galloway Burke, and Colin Cannonier. "African American Women Leadership across Contexts." *Journal of Management History* 26.3 (2020): 353–376.

NOTES

9. Shannon Sales, Monica Galloway Burke, and Colin Cannonier. "African American Women Leadership across Contexts." *Journal of Management History* 26.3 (2020): 353–376.
10. Megan Bailey. "Between Two Worlds: Black Women and the Fight for Voting Rights." National Park Service. Last updated October 9, 2020. https://www.nps.gov/articles/black-women-and-the-fight-for-voting-rights.htm.
11. Olivia B. Waxman. "Stacey Abrams and Other Georgia Organizers Are Part of a Long—But Often Overlooked—Tradition of Black Women Working for the Vote." *Time*. January 8, 2021. https://time.com/5909556/stacey-abrams-history-black-women-voting/.
12. Olivia B. Waxman. "Stacey Abrams and Other Georgia Organizers Are Part of a Long—But Often Overlooked—Tradition of Black Women Working for the Vote." *Time*. January 8, 2021. https://time.com/5909556/stacey-abrams-history-black-women-voting/.
13. Olivia B. Waxman. "Stacey Abrams and Other Georgia Organizers Are Part of a Long—But Often Overlooked—Tradition of Black Women Working for the Vote." *Time*. January 8, 2021. https://time.com/5909556/stacey-abrams-history-black-women-voting/.
14. Jessica Washington and Tiffany Arnold. " 'Whatever It Takes': How Black Women Fought to Mobilize America's Voters." *The Guardian*. November 12, 2020. https://www.theguardian.com/us-news/2020/nov/12/black-women-voters-mobilize-georgia-elections.

Chapter 2: How Not to Be Erased

1. Gillian Greensite. "History of the Rape Crisis Movement." CALCASA: Support for Survivors: Training for Sexual Assault Counselors. 2003. https://www.valor.us/2009/11/01/history-of-the-rape-crisis-movement.

NOTES

2. Treva Lindsey. "Black Women Have Consistently Been Trailblazers for Social Change. Why Are They So Often Relegated to the Margins?" *Time*. July 22, 2020. https://time.com/5869662/black-women-social-change/.
3. Stewart M. Coles and Josh Pasek. "Black Women Often Ignored by Social Justice Movements" (press release). American Psychological Society. July 13, 2020. https://www.apa.org/news/press/releases/2020/07/black-women-social-justice.
4. Parul Sehgal. "Fighting 'Erasure.'" *New York Times*. February 2, 2016. https://www.nytimes.com/2016/02/07/magazine/the-painful-consequences-of-erasure.html.
5. Parul Sehgal. "Fighting 'Erasure.'" *New York Times*. February 2, 2016. https://www.nytimes.com/2016/02/07/magazine/the-painful-consequences-of-erasure.html.
6. Maya Finoh and Jasmine Sankofa. "The Legal System Has Failed Black Girls, Women, and Non-Binary Survivors of Violence." ACLU. January 28, 2019. https://www.aclu.org/blog/racial-justice/race-and-criminal-justice/legal-system-has-failed-black-girls-women-and-non.
7. Maya Finoh and Jasmine Sankofa. "The Legal System Has Failed Black Girls, Women, and Non-Binary Survivors of Violence." ACLU. January 28, 2019. https://www.aclu.org/blog/racial-justice/race-and-criminal-justice/legal-system-has-failed-black-girls-women-and-non.
8. "5 Facts about Black Women in the Labor Force." *U.S. Department of Labor Blog*. August 3, 2021. https://blog.dol.gov/2021/08/03/5-facts-about-black-women-in-the-labor-force.
9. Sarah Jane Glynn. "Breadwinning Mothers Continue to Be the U.S. Norm." Center for American Progress. May 10, 2019. https://www.americanprogress.org/issues/women/reports/2019/05/10/469739/breadwinning-mothers-continue-u-s-norm/.

NOTES

10. Stacia Kirby. "A Historic Number of Black Women Own Small Businesses in the US." Cision PRWeb. May 15, 2022. https://www.prweb.com/releases/a_historic_number_of_black_women_own_small_businesses_in_the_us_finds_annual_small_business_trends_report_released_by_guidant_financial/prweb17758718.htm.
11. Andre Dua et al. "COVID-19's Effect on Minority-Owned Small Businesses in the United States." McKinsey & Company. https://www.mckinsey.com/industries/public-and-social-sector/our-insights/covid-19s-effect-on-minority-owned-small-businesses-in-the-united-states.
12. Trena Easley Armstrong. "The Hidden Help: Black Domestic Workers in the Civil Rights Movement." ThinkIR: The University of Louisville's Institutional Repository. November 2012. https://ir.library.louisville.edu/cgi/viewcontent.cgi?article=1045&context=etd.

Chapter 3: Walking the Tightrope

1. "Statement from Vice President Kamala Harris on the Senate Vote on Voting Rights." White House Briefing Room. January 19, 2022. https://www.whitehouse.gov/briefing-room/statements-releases/2022/01/19/statement-from-vice-president-kamala-harris-on-the-senate-vote-on-voting-rights/.
2. Andie Kramer. "Recognizing Workplace Challenges Faced by Black Women Leaders." *Forbes*. January 7, 2020. https://www.forbes.com/sites/andiekramer/2020/01/07/recognizing-workplace-challenges-faced-by-black-women-leaders/?sh=4f987f1653e3.
3. Andie Kramer. "Recognizing Workplace Challenges Faced by Black Women Leaders." *Forbes*. January 7, 2020. https://www.forbes.com/sites/andiekramer/2020/01/07/recognizing-workplace-challenges-faced-by-black-women-leaders/?sh=4f987f1653e3.

NOTES

4. Andie Kramer. "Recognizing Workplace Challenges Faced by Black Women Leaders." *Forbes*. January 7, 2020. https://www.forbes.com/sites/andiekramer/2020/01/07/recognizing-workplace-challenges-faced-by-black-women-leaders/?sh=4f987f1653e3.
5. Ashleigh Shelby Rosette and Robert W. Livingston. "Failure Is Not an Option for Black Women: Effects of Organizational Performance on Leaders with Single versus Dual-Subordinate Identities." *Journal of Experimental Social Psychology* 48.5 (2012): 1162–1167.
6. Ashleigh Shelby Rosette and Robert W. Livingston. "Failure Is Not an Option for Black Women: Effects of Organizational Performance on Leaders with Single versus Dual-Subordinate Identities." *Journal of Experimental Social Psychology* 48.5 (2012): 1162–1167.
7. Anna Zarra Aldrich. "Leading While Black, the Experience of Black Female Principals." *UCONN Today*. May 12, 2019. https://today.uconn.edu/2019/03/spencer-foundation/.
8. Marianne Cooper. "Women Leaders Took on Even More Invisible Work during the Pandemic." *Harvard Business Review*. October 13, 2021. https://hbr.org/2021/10/research-women-took-on-even-more-invisible-work-during-the-pandemic.

Chapter 4: Our Fight

1. Scott Winship et al. "Long Shadows: The Black-White Gap in Multigenerational Poverty." Brookings. June 10, 2021. https://www.brookings.edu/research/long-shadows-the-black-white-gap-in-multigenerational-poverty/.
2. Maria Guerra. "Fact Sheet: The State of African American Women in the United States." Center for American Progress. November 7,

NOTES

      2013. https://www.americanprogress.org/article/fact-sheet-the-state-of-african-american-women-in-the-united-states/.

3. Maria Guerra. "Fact Sheet: The State of African American Women in the United States." Center for American Progress. November 7, 2013. https://www.americanprogress.org/article/fact-sheet-the-state-of-african-american-women-in-the-united-states/.

4. Maria Guerra. "Fact Sheet: The State of African American Women in the United States." Center for American Progress. November 7, 2013. https://www.americanprogress.org/article/fact-sheet-the-state-of-african-american-women-in-the-united-states/.

5. Piraye Beim. "The Disparities in Healthcare for Black Women." Endometriosis Foundation of America. June 6, 2020. https://www.endofound.org/the-disparities-in-healthcare-for-black-women.

6. Veronica Zaragovia "Trying to Avoid Racist Health Care, Black Women Seek Out Black Obstetricians." NPR. May 28, 2021. https://www.npr.org/sections/health-shots/2021/05/28/996603360/trying-to-avoid-racist-health-care-black-women-seek-out-black-obstetricians.

7. Mathilde Roux. "5 Facts about Black Women in the Labor Force." *U.S. Department of Labor Blog*. August 3, 2021. https://blog.dol.gov/2021/08/03/5-facts-about-black-women-in-the-labor-force.

8. Mathilde Roux. "5 Facts about Black Women in the Labor Force." *U.S. Department of Labor Blog*. August 3, 2021. https://blog.dol.gov/2021/08/03/5-facts-about-black-women-in-the-labor-force.

9. Rob Picheta. "Black Newborns More Likely to Die When Looked After by White Doctors." CNN. August 20, 2020. https://www.cnn.com/2020/08/18/health/black-babies-mortality-rate-doctors-study-wellness-scli-intl/index.html.

NOTES

Chapter 5: Our Sacrifice

1. Sonia Saraiya. "Viola Davis: 'My Entire Life Has Been a Protest'" *Vanity Fair*. July 14, 2020. https://www.vanityfair.com/hollywood/2020/07/cover-story-viola-davis.
2. Aquilah Jourdain. "Suffering and the Black Female Narrative in the Twentieth Century" (thesis). CUNY City College. 2019. https://academicworks.cuny.edu/cc_etds_theses/791/.
3. Kara Manke. "How the 'Strong Black Woman' Identity Both Helps and Hurts." *Greater Good Magazine*. December 5, 2019. https://greatergood.berkeley.edu/article/item/how_the_strong_black_woman_identity_both_helps_and_hurts.
4. Kara Manke. "How the 'Strong Black Woman' Identity Both Helps and Hurts." *Greater Good Magazine*. December 5, 2019. https://greatergood.berkeley.edu/article/item/how_the_strong_black_woman_identity_both_helps_and_hurts.

Chapter 7: "A Little Child Shall Lead Them"

1. Isaiah 11:6 (English Standard Version).
2. Erica L. Green, Mark Walker, and Eliza Shapiro. "'A Battle for the Souls of Black Girls.'" *New York Times*. October 1, 2020. https://www.nytimes.com/2020/10/01/us/politics/black-girls-school-discipline.html.
3. Jacquelyne Germain. "Simone Biles, Sha'Carri Richardson, and How the Olympics Failed Black Women." ACLU. August 13, 2021. https://www.aclu.org/news/racial-justice/simone-biles-shacarri-richardson-and-how-the-olympics-failed-black-women.

NOTES

4. "The Case for Focusing on Black Girls' Mental Health." National Black Women's Justice Institute. June 7, 2021. https://www.nbwji.org/post/the-case-for-focusing-on-black-girls-mental-health.

5. "The Case for Focusing on Black Girls' Mental Health." National Black Women's Justice Institute. June 7, 2021. https://www.nbwji.org/post/the-case-for-focusing-on-black-girls-mental-health.

6. "The Case for Focusing on Black Girls' Mental Health." National Black Women's Justice Institute. June 7, 2021. https://www.nbwji.org/post/the-case-for-focusing-on-black-girls-mental-health.

About the Author

White House Correspondent April Ryan has a unique vantage point as the only Black female reporter covering urban issues from the White House—a position she has held for twenty-five years, since the Clinton era. This position has afforded her unusual insight into the racial sensitivities, issues, and attendant political struggles of our nation's presidents.

April can be seen almost daily on CNN as a political analyst and is the Washington, DC, bureau chief on TheGrio. She has been featured in *Essence*, *Vogue*, *Cosmopolitan*, and *Elle* magazines, to name a few. She also has served on the board of the prestigious White House Correspondents' Association, and is one of only three African Americans in the association's over one-hundred-year history to do so. She is also an esteemed member of the National Press Club. In 2015, April was nominated for an NAACP Image Award (Outstanding Literary Work—Debut Author) for her first book. In 2016, she received the National Council of Negro Women's Dr. Mary McLeod Bethune Trailblazer Award.

ABOUT THE AUTHOR

In 2019, April became an honorary member of Delta Sigma Theta Sorority, and was recognized as the Freedom of the Press Award winner by the Reporters Committee for Freedom of the Press. She was nominated in 2021 for the NAACP Image Award for Social Justice Impact. A Baltimore native and a Morgan State University graduate, April gives back to this community by serving as a mentor to aspiring journalists and assisting with developing "up-and-coming" broadcasters. She considers her greatest life's work raising her two daughters, Ryan and Grace, who are phenomenal young women.

April is the author of the award-winning book *The Presidency in Black and White: My Close-Up View of Three Presidents and Race in America*; *At Mama's Knee: Mothers and Race in Black and White*, in which she looks at race relations through the lessons and wisdom that mothers have given their children; and *Under Fire: Reporting from the Front Lines of the Trump White House*.